THE
DINGHY-OWNER'S
HANDBOOK

The Dinghy-Owner's Handbook

DAVE JENKINS

Illustrations by
the author

HOLLIS & CARTER
LONDON   SYDNEY
TORONTO

Paperback edition I S B N 0 370 10346 7
Hard-cover edition I S B N 0 370 10243 6
Printed in Great Britain for
Hollis & Carter
an associate company of
The Bodley Head Ltd
9 Bow Street, London W C 2 E 7 A L
by Northumberland Press Ltd, Gateshead
Set in Linotype Baskerville
*First published as a paperback and
simultaneously as a hard-cover edition 1975*

CONTENTS

1

Just for the Fun of It, 9
*So Many Kinds of Boats, 10 · Where Shall We Sail?,
12 · Why Join a Club?, 16 · The Lone Sailors, 18 ·
Learning to Sail, 19 · Reckoning the Cost, 22*

2

Casting an Eye Over Dinghies, 27
*Materials and Construction Methods, 27 · Hull
Shape and Purpose, 36 · Decking and Layout, 40 ·
The Rig, 43 · Single-Sail Rigs, 47 · Other Rigs,
48 · The Mast as Part of the Rig, 49*

3

The Choice of a Dinghy, 51
*Trial Sails, 51 · A General Idea of What You Want,
52 · One-Design and Restricted Classes, 61 · Local
One-Designs, 62 · Prices—The Crunch, 64 · The
Second-Hand Boat, 68 · Class Measurement Certifi-
cate and Ownership, 69*

4

Sails, 71
*Their Setting, Trimming and Control, 71 · Rigging
Adjustment and Tuning, 91 · Wire Rope Breaking*

*Stresses, 93 · Rod Wire Breaking Stresses, 95
Natural and Synthetic Rope, 95 · Whipping, 97 ·
Rope Splicing, 97 · Knots, Bends and Hitches, 97 ·
Suggested Choice of Ropes and Sizes, 99 · Sailcloth
weight, 100*

5

Masts, Spars and Ancillary Equipment, 101
*Masts and Spars, 101 · Shackles and Blocks, 111 ·
Specialised Blocks, 116 · Mainsheet Systems, 118*

6

Lateral Resistance and Steering, 123
*The Centreboard, 123 · Steering, 130 · Heeling and
Righting Levers, 138 · Reducing the Heeling Mo-
ment, 143 · Reefing, 144*

7

Capsize and Self-Rescue, 149
*Clothing for Sailing, 149 · Personal Buoyancy, 156 ·
Getting Cold, 161 · Buoyancy for the Boat, 163 ·
Capsize and Self-Rescue, 168 · Draining the Boat,
174 · Returning to Shore, 178 · Calling for
Assistance, 182*

8

Wind, Water and Helmsmanship, 185
*Wind and Weather, 185 · Current and Tidal Flow,
195 · Waves, 200 · Special Sailing Techniques, 203 ·
Anchoring and Mooring, 209*

9

Racing Your Boat, 214

Racing Your Boat, 214 · Race Procedure, 216 · When Boats Meet, 219 · Rounding and Giving 'Room' at Marks, 222 · Beginning to Race, 226 · Spinnaker, the Downwind Sail, 235 · Fair and Foul Means of Propulsion, 241 · The Finish, 242 · Flag Signals, 242

10

Handling Your Boat Ashore, 244

Carrying Your Dinghy, 244 · The Dinghy Park, 260 · The Boat Cover, 261 · Laying-Up, 262 · Resinglass Maintenance, 264 · Wooden Boat Maintenance, 265 · Application of Paint and Varnish, 270

Appendix:

Useful References, 273

Diagrams and Tables for Easy Reference, 275

Index, 277

I

Just for the Fun of It

At the moment, having a dinghy of your own may be an unfulfilled wish. This may be due to lack of resources or to hesitation before taking a major step—one apparently teeming with unknown difficulties and responsibilities. Then again, you may have neither the desire nor facilities for boat ownership but the opportunities to sail in other people's. Perhaps, happily, you are already the proud possessor of a craft. Whichever is the case this book is intended to give help and encouragement.

Enthusiasm for sailing is an ingredient you probably already have in plenty and, although it can be stimulated by book-reading, the physical act of sailing and the contact of boats does this so much better. Sailing can never be dull, even on those lazy days, when the burgee at the masthead can scarcely raise a flutter and the lapping water under the forefoot seems to raise a chuckle as a stronger breath urges the boat forward; or on those memorable summer days, when the dinghy crunches lightly on to the picnic beach and thoughts are first given to securing the boat before lightening the lunch basket or repairing to the local inn.

Then there are the boisterous days, when the boat is full of life and the air full of spray as she scutters away on a planing reach with foaming bows and a creaming wake; days when you must exercise judgement and seamanship by deciding just what kind of sailing you, your crew, and your boat should take—whether to go out fully rigged into an exhilarating welter of wind and take the risks involved, or whether to reef the sail and be safer. Racing often overrides better judgements in this, but not all races are won with full sail.

And then there are the days which are not so good, when you are caught out in sheeting rain and the 'openness' of an open boat becomes fully apparent; or cold, misty days when the spars and rigging 'weep' water and the ropes and sheets numb the fingers. Yet even on such days there is compensation in the peace and escape from the workaday world, and sudden enjoyments, heightened by their unexpectedness: the flash of sunshine after rain which transforms the water into a sheet of shimmering silver, or the close-up view of a grebe or a cormorant afforded by a shrouded approach through the mist.

SO MANY KINDS OF BOATS

Today's newcomer, and sometimes even an experienced boatman, may query the need for the bewildering variety of boats that may be seen around—perhaps with justification when there seem to be only slight differences. There are historical reasons. The early boats used entirely for pleasure were converted work boats and therefore built to stand up to a lot of knocking about and to all-the-year use with minimum maintenance. But the new breed of owner took his boat out of the water during the winter months, had more time and money to maintain her, and neither expected her to withstand rough treatment nor to be a load carrier. For all these reasons boats were built lighter, a trend which was given added incentive as yacht and dinghy racing became a popular pastime and speed and responsiveness in a boat became the important virtues.

The availability of boat-building materials has been a great influence on boat design. Traditional building, using strips of solid wood produced rounded hulls. Marine ply—used most effectively in panels—gave rise to chine boats, with angular sections, then, with the coming of plastics, the return to compound curvature in hulls was made not only possible but desirable for strength.

Sailing waters vary from open, exposed coastal, with deep or shallow water, to narrow rivers or tiny lakes. These conditions have an enormous bearing on the kind of boat which is successfully sailed. A large boat may be preferred not because she carries more crew but because she will look after her occupants better in rougher water and in more comfort than will a smaller one.

The above considerations result in 'types' of boat and if a number of boats are built of a type, and are given similar dimensions, then this is described as a 'class' and usually given a class name—Enterprise and Lymington Scow are examples. Classes will be given a fuller review in Chapter 3.

The proliferation of dinghy classes, some with not dissimilar measurements, is largely due to many designers and builders, quite naturally, desiring a share of the market. This is a 'richness' it may be as well to appreciate while it lasts. There could be a parallel in the motor-car industry which, in its infancy, provided designing opportunities for every small-town garage mechanic. Now we are 'stuck' with a few producers and a limited choice. With mass-production methods taking over the boat-building industry future choice could be limited in the same way. Long may the 'small' boatbuilder and designer live!

Class dinghy

LOA 13ft 3in (4·03m)
Beam 5ft 3in (1·59m)
Weight 220lb (100kg)
Sail Area 113 sqft
(10·5 sq m)

SAIL MARK **E**

The Enterprise one-design is ideal for racing and suitable for limited cruising with a reduced rig

WHERE SHALL WE SAIL?

There are so many personal aspects to this that it is possible only to chat for a while about the pros and cons of various types of sailing area and then return the question firmly back to you.

Personally, I have always found that transport to and from the area has made most of the decisions for me. Living near London in my 'teens with no other form of transport than a bicycle I could manage the thirty-odd miles to the Thames. With a car, in my late twenties I was hurtling down to the East Coast every weekend until, in my mid-thirties and with increasing road congestion, I began to wonder if it was worth all that expenditure of nerves, rubber and petrol for the limited sailing time that tides allow. So a local flooded gravel pit, with its drawbacks of limited size and 'enforced' class racing, became more attractive until these recent years when, 'knocking-on fifty' I moved to North Norfolk. Now transport is little problem and I can almost walk to the local harbour and am restricted only by tidal flow.

These personal notes are given only to show how one aspect—transport—can have a bearing on choice. Another consideration is the amount of discomfort you, (and your family, if sailing is to be a combined venture), are prepared to tolerate.

Most dinghies are open boats—although a cuddy or an ample foredeck is fitted to some to protect the more fragile crew members—and exposure to the weather is something one has just got to put up with. For coastal sailing there is more 'putting up with it' to be tolerated than inland. Even estuaries and wide rivers near the sea can provide waves into which your boat will thump to cause 'green-uns' to come aboard and douse the occupants from head to foot. Launching and landing may well be more hazardous, with waves trying to turn the boat over, and it will probably also be necessary for one member to wade waist-high at these times. Protective dress is essential and keeps down the discomfort, but it cannot eliminate it.

The sailing season tends to be shorter on the coast than on inland waters because of these tougher conditions— probably May till September instead of all the year round, if that is your inclination. Even during this shorter season there will be many more days on the coast than inland when strong winds make sailing impossible as the 'sea breezes' add their weight to the normal gradient winds.

GP 14 (One-design Class)
LOA 14 ft (4·27 m)
Beam 5 ft (1·52 m)
Weight 285 lb (130 kg)
Sail Area 102 sq ft
(9·5 sq m)

Estuary sailing

Estuaries usually provide open water sailing, tidal limitations with swift currents, conditions that are often rough and wet feet for launching and landing

Yet there are so many attractions to sea and estuary sailing that many will consider no other. For those who have tried the confines of inland waters the room to enable long tacks and courses to be made gives great satisfaction. Wind is usually truer and steadier, which compensates for its greater strength, and there is less likelihood of shore obstructions blocking it and causing mischievous eddies. There can also be environmental attractions, such as a nearby beach which could keep the 'less interested in sailing' members of the family happy.

The safety aspect of a sailing venue demands some appreciation. Undoubtedly the sea is less safe than inland water as, in the event of a capsize or any other incapacity, there is no safe shoreline in whichever direction the wind chooses to blow you, as there is with inland water. It could also be said that the sea is no place on which to learn to

River sailing

MERLIN ROCKET
(Restricted Class)
LOA 14ft (4·27m)
Beam 5ft 1in (1·54m)
Weight 200 lb (91 kg)
Sail Area 105 sq ft
(9·8 sq m)

A river of moderate width, like the Thames, provides safe racing and cruising. Currents are normally weak and winds fickle and variable

sail—with justification if you set out for the first time with the wind blowing off the land. But with some expert guidance, with the right weather conditions and with a lot of thought put into the preparations, it is as good a place to begin as any.

Inland water, with the exceptions of the larger lakes in these islands, (and if you have one of these within easy reach you are lucky indeed) is always confined and year by year is becoming more and more congested. Rivers, which used to give such excellent sailing, now have their widths reduced by motor cruisers lining the banks and the fairways impeded by marine traffic. But rivers are fine waters for learning to sail on, as they are comparatively safe and have the added interest and challenge that a moving body of water always brings.

It is probably due, in some measure, to the congestion of the rivers that reservoir and flooded gravel pit sailing has now become so attractive to dinghy sailors. These waters vary in size from the several square miles of such reservoirs as Grafham to the twenty or thirty acres of some of the smaller gravel pits. Yet all of them offer fine open-water sailing and excellent learning facilities. The normal 'hazards' that one just learns to accept are shallow water and fluky winds.

The man-made draw-backs it may be more difficult to accept are that usually these pieces of water are leased only to clubs so that there is no provision for the 'casual' sailor. Then, having decided to join, the new member finds that he cannot sail during times allotted to racing, unless he takes part.

WHY JOIN A CLUB?

If, as exemplified in the last sentence, a club restricts a member from doing what he wants to do, this may seem a good reason for not joining. But there are a host of advantages and the restrictions, although irksome to some, should be generally agreed to favour the majority of the membership. In the previously stated case, the waters are probably too small to have jilling sailors around during racing while, for another example where the member is restricted in his choice of class boat to three or four, this is the only way to ensure that the majority can enjoy the best kind of racing—class racing, when all the boats are of the same type, rather than handicap racing, when boats of various types attempt to race.

Most people join clubs for the facilities they offer. For gravel pits and reservoirs, sailing water must be counted as one but easy launching, changing and washing amenities, storage of sails and other 'clobber', refreshment rooms, dinghy and car parking are others in a list which could be endless.

Some people join clubs for the racing, but probably not their first club. Racing is a pastime into which they are drawn by curiosity and then become addicted to, rather than having it as a definite objective from the start of their sailing lives. Many clubs prosper because of their racing programme and draw sailors from other clubs by it, and by maintaining enthusiastic fleets in certain classes.

Others are attracted to clubs for their social life. This is not just the desire to sail in company, for most clubs can offer much more than that. At the informal level there are the chats in the dinghy park and in the bar. Then there are the 'socials', club dances and dinners and unrelated items like Firework and Hallowe'en Nights or organised visits to the theatre. Then the widening circle includes reciprocal visits between clubs, which gives members the chance to sail on other waters and to meet sailing types from around the country.

A few people join clubs for the status it brings. This sounds like snobbery and very often this is just the word for it. But if it helps some sailors, when sailing away from home waters, to behave better out of loyalty to their club, that cannot be a bad thing either. A club membership card is a form of credential, for no reputable club will tolerate any 'black sheep' and card-carrying members usually find a universal camaraderie at other clubs they may visit on holiday or when travelling around, and may be treated as guests or be allowed temporary membership if they should desire it.

THE LONE SAILORS

Well, of course, they sail round the world don't they? Actually this was not quite the image I wished to evoke by that heading or even to point out that we are all either 'clubbable' or 'non-clubbable'. But many of us, and many families find it better or more convenient, to sail alone —and why not?

Despite the impression given by the yachting press, countless more dinghy sailors do not belong to clubs than do. They range from the real 'loner', the chap with his clinker boat down some muddy creek, approachable only with thigh-waders, who spends the whole weekend in her and sleeps under a tarpaulin, to the family who carry the dinghy on the roof rack down to the sea 'for the kids to play around in' and who could be the subjects for a lifeboat rescue when they get into trouble.

This is not the inevitable, dire result of all lone or family sailing and I am not decrying it; but I am just pointing out that to sail not in company, or without supervision from afloat or ashore, brings added responsibility to the participants for their own safety. One thing that can be said about clubs, especially on the coast, is that sailing is invariably supervised, with rescue boats standing by, and that racing is restricted or cancelled in bad or predicted bad weather. For solo sailors the responsibility is theirs.

LEARNING TO SAIL

One thing that clubs do not do is to teach newcomers to sail. Nevertheless, membership does not depend on an ability to sail, so if you are a newcomer and decide to join notwithstanding, make yourself known by hanging around and helping the boats in and out. You will quickly inveigle yourself into crewing jobs in other people's boats and pick up your sailing as you go along. In fact, crewing for a good helmsman is one of the best ways to learn one's sailing—getting started is the trouble.

With your own boat in a club the policy could be a mixture of the above and inviting more experienced members to sail with you. A declared lack of expertise is never looked down on and club people are usually ready to help. What is less readily accepted is a declaration of experience one does not possess. This could lead to a helmsman taking out a beginner on a difficult, windy day, with discomforting results for both.

With a boat and no desire to join a club, at least until you have acquired some competence with her, the simple solution is to garner as much 'theory' from books, magazines and courses as you can, then take your boat to the water and just put theory into practice. With care, attention to weather and sea conditions, and an alert mind for other complications and dangers, there is a lot to be said for this method. But it might be better to take along a competent friend as well. There must be a slight reservation about the friend, for bad guidance is worse than none at all.

One way to get competent tuition is to take a course at a sailing school. These schools were springing up like mushrooms until 1968 when two bodies—the RYA (Royal Yachting Association) and the NSSA (National Schools Sailing Association) co-operated to rationalise sail and small-boat instruction in this country and to raise and maintain the standards of training establishments and instructors. The result was the RYA National Proficiency Certificate Scheme and the National Coaching Scheme.

Full details of these schemes is available from the RYA (booklet G4). Actually, the National Coaching Scheme is for instructors, intending instructors and race officials at present, but anyone can qualify for the National Proficiency Scheme. The idea is to have Dayboat Certificates in three grades—Elementary, Intermediate and Advanced—issued to successful applicants for the theoretical and practical tests given at authorised test centres. These are mostly sailing schools, who run training courses, but there are also a few sailing clubs with training sub-committees.

Of course there is nothing to stop you taking a course at a sailing school without bothering with the test and certificate. These are commercially run establishments and they supply all the ingredients for a good holiday—accommodation, meals, social life, and get you started the right way in sailing into the bargain—but it would be wise to pick one of the RYA-recognised schools just the same.

Children enjoy a special provision under the Scheme; they are looked after by the NSSA and an enquiry to them will elicit full details and the address of the nearest Association to your area.

Supplementary to the National Scheme many Local Authorities and Education Committees organise sailing activities as part of school curriculum. There may be schools that are a little lax in making the best of these for their pupils so it could be as well, if you have interested children, to make enquiries. An extension of these facilities to adults, as further education classes, is also carried out in many areas.

For a family starting sailing together, getting the children trained first under one of these schemes, for their expertise then to be disseminated to the rest, is not such a bad idea. They are quick learners and are bound to appreciate being the 'wise ones' for a change.

International Cadet Class: I can hardly, in the context of children learning to sail, leave out some reference to the Cadets, which have taught so many of our younger generation helmsmen to sail. Many sailing clubs run Cadet sections which accept children between the ages of eight and seventeen. They sail the well-known, hard chine, $10\frac{1}{2}$ ft dinghy, also called Cadet, and are fully supervised by interested parents during the racing and other associated activities.

Instruction is given not only about racing skills but on all aspects of boating, in a strong and lively club. The children learn from their mistakes—in a safe boat so there is no damage done—from each other and from ex-Cadets, boys and girls, who frequently continue to take an interest after they have reached 'retiring age'.

Boats are owned by the members or their parents but there is no prerequisite of owning a boat for joining. A boatless member usually finds a place in a boat, crewing, so that expenses are limited to club fees.

Junior dinghy

C
SAIL MARK

LOA 10ft 6 in (3·22m)
Beam 4ft 2in (1·27m)
Weight 150lb (68 kg)
Sail Area 56 sq ft
(5·2 sqm)

The International Cadet is a one-design, class dinghy. When young children start they often look too small for the boat; eventually they look too big!

RECKONING THE COST

They used to say years ago that—'if it's your hobby, you don't count the cost'. Perhaps there is still some truth in it but I suspect the phrase was coined when hobbies—stamp collecting, model aircraft construction, photography—had a more moderate scale of prices, were less sophisticated and needed less expensive accessories than they do now.

Nowadays, most people must count the cost and trim their financial sails as well as their Terylene ones.

Obviously the biggest single outlay will be on the boat herself. Boats are to be discussed in some detail in the next chapter and it will suffice to say here that, in spite of the high cost of new boats, they hold their value sufficiently well to have a reasonable resale value, so much of that outlay may be redeemed. Purchase of a second-hand boat by the astute, followed by careful renovation and maintenance, can even result in a slight profit on the resale.

Sailing without one's own boat, by joining a club and crewing for someone, is feasible for one person but for a couple or a family this solution is not on. The only alternative is to hire or charter. This can be by the day, or part of a day, or by the week. This may appear an expensive way to sail but it all depends how much sailing you do. I hired for years before I could afford my own boat and eighteen shillings for half a day's sailing every weekend or so seemed to be pocket-money well spent. Hirers can mentally debit from their costs the boat-owner's expenses of maintenance, depreciation and insurance and the trouble of storage, towing and rigging. To set against that there are the disappointments in the availability of a boat and often her condition.

To the depreciation factor of boat ownership one must add the cost of annual maintenance. With breakages, periodic renewal of items which wear out, including sails, this can vary from a modest outlay for a resinglass boat (which despite no-maintenance claims may still require her hull painted after a year or so) to quite a substantial one for the keenly raced wooden boat (which requires full painting and varnishing and may be fitted with new sails every two years together with new sheets, halyards and buoyancy bags).

Then transport. While a small boat can be carried by road on the top of a car, most two-man dinghies require a road trailer. Little should be needed in the way of maintenance but in the annual budget a few pounds must be earmarked for depreciation and the renewal of tyres. A launching trolley is usually needed as well.

Sailing water is seldom free. Even on the coast there may be harbour, foreshore or car park charges. Inland, in addition to the fees of the sailing club which it may be necessary to join to sail on certain waters, there could be Local Water Authority or River Conservancy tolls to consider.

Storage of the boat is another thing. If this is to be at home and the boat trailed to the water every time she is sailed then there is no problem, but storing her during the season in the club or municipal dinghy park costs money. During the winter months she should be stored under cover and if you have not got the accommodation this puts another item on the bill.

Insurance is a 'must' for boat owners, not only for the protection of this valuable asset and her trailer but to cover the third-party claims that could occur and for which you could be made liable—either on the water or on the road. Most comprehensive motor insurances will cover the third-party risks of towing a boat but not the boat herself. Marine insurance, with only a low premium, will take care of the 'gaps' in the cover.

The most personal cover is clothing and for dinghy sailing you must have a good wardrobe of waterproofs, non-slip rubber shoes and a lifejacket or buoyancy-aid. Initial kitting out could run up a sizeable bill after which a small annual replacement cost should suffice.

I have deliberately not mentioned figures: in a world of rising prices they would probably be too far on the low side once they got into print. The object of this section has been simply to show the newcomer to dinghy sailing what items to bear in mind when budgeting. As an attempt to collate these items I have set them down in the table overleaf, as a footnote, with the basic assumption that one person with a boat, living 50 miles from the coast, and 15 miles from the alternative inland sailing water, is comparing the relative costs of sailing in the two areas. In spite of the lower club fees, which is usual for coastal clubs, and assumed free dinghy parking, the distance makes his sailing on the coast cost more than half as much again as sailing at his local club's water. Of course if he had lived nearer the coast the result would have been different. We have all got to apply our own circumstances and permutation of the other variables, in preparing such a table of estimates. For example, the club fees would be double, or more, for a couple or for family membership.

Remember too, that if such a list is prepared to estimate the cost of getting started in sailing, there may be an entrance fee for club membership the first year, that the personal clothing entry will be higher and that the major outlay of the initial cost of the boat and trailer must be set against the entries for maintenance and depreciation of them.

DINGHY OWNER'S ANNUAL OUTLAY

The figures are purely for purposes of comparison. The reader is invited to substitute corrected figures for his own sailing requirements.

£

	Coastal	Inland
Club fees (one member)	6	18
Transport (at 5p per mile)		
30 trips of 100 miles	150	
30 trips of 30 miles		45
Car parking	6	2
Maintenance—boat, boatcover,		
sails, trailer and trolley	40	30
Depreciation of same	100	60
Boat storage or parking		
summer	–	12
winter	5	5
Insurance (comprehensive)		
of boat, trailer, trolley	6	6
Personal clothing and		
buoyancy equipment	17	17
Local Water Authority dues	–	5
(there could be coastal dues)		
	330	200

Casting an Eye Over Dinghies

Before getting down to specific boats and the agonising question of an individual choice, I believe it is a good idea to review dinghy design in a general way.

MATERIALS AND CONSTRUCTION METHODS

Wooden Clinker: The traditional building method. The boat is built upside-down on moulds shaped to the sections of the hull. They, and the shaped stem, hog and transom are strutted temporarily into position while the planking, which must be 'tailored' to accommodate the curvature, is glued, in an overlapping sequence from the keel to the gunwale, to consolidate the structure. The 'shell' is then removed from the moulds for the internal joinery to be done.

A more heavily built boat will have the 'lands' (overlaps) through-fastened with copper rivets, and narrow ribs, steamed to shape, fixed inside at, say, 6-in intervals. The planking may be of solid wood (larch or mahogany) or of marine plywood (in the case of racing dinghies) and number from about a dozen narrow ones (for the solidly built 'knockabout') to only four or five (for a lightly built racing shell).

Virtues: An immensely strong construction but one for the professional or skilled amateur. Makes up into a

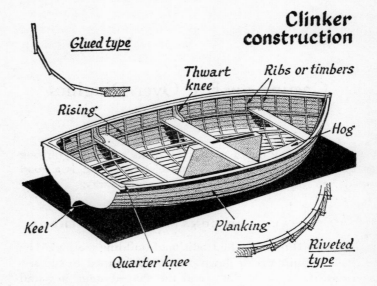

Clinker construction

Glued type

Thwart knee

Ribs or timbers

Rising

Hog

Keel

Planking

Quarter knee

Riveted type

round-bilged, and usually very seaworthy hull. The lands help to keep down the spray.

Drawbacks: Cleaning-out, especially when fitted with ribs, is more troublesome, as is maintenance. The problem of leaking, which used to be a bugbear when boats had to 'take-up' with moisture to make them tight, is seldom met with today with modern synthetic resin glues. With more 'wetted area' it has a 'slower' performance than comparable 'smooth hulled' boats.

Moulded Plywood: There are 'hot' and 'cold' moulded hulls, the difference being only in the 'curing' of the glue. Again, the boat is built upside-down over moulds, but this time more complete ones with longitudinal laths covering the section moulds and fairing into the stem, hog and transom. Thin wood veneers (mahogany, sapele or similar hardwood) about 2 inches wide are stapled to the frame, diagonally from gunwale to keel. At first only the alternate strips are put on, then the spaces filled by others 'tailored' as required. Next, as the temporary staples

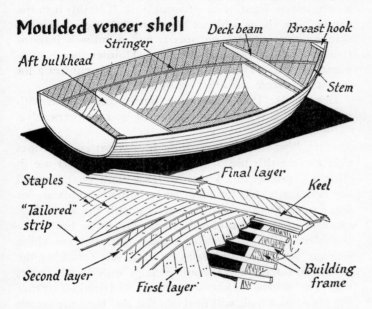

Moulded veneer shell

Deck beam — Breast hook — Stringer — Aft bulkhead — Stem — Staples — Final layer — Keel — "Tailored" strip — Second layer — First layer — Building frame

are withdrawn, another layer of veneers is stapled and glued over the first layer, diagonally opposite to them. The build-up is completed by a third layer, usually in a fore-and-aft direction.

For cold moulding the glue cures at normal temperatures and the hull is removed from the mould for finishing, but for hot moulding, which is done by the firm of Fairey Marine who alone have the necessary ovens and equipment, a different type of glue is used which cures only with heat. In this case the whole shell is placed inside a rubber bag from which the air is then evacuated. This applies atmospheric pressure to squeeze the laminations while the hull is 'baked' in a large oven to cure the glue.

Virtues: For both hot and cold methods this results in a very strong, light, homogeneous hull. It builds into the round bilge shape without coercion and offers a smooth skin on the outside (for speed) and a smooth interior (which makes for ease of cleaning). Maintenance is comparatively easy. The cold moulded method requires a lot of patience from the amateur builder.

Drawbacks: Practically none, but in competition with resinglass this type of building must be rated expensive.

Marine Plywood, Chine: The hard corners or chines differentiate this use of marine plywood from its use in clinker construction. There are two main methods and the first again calls for inverted building. Moulds are set up, shaped to the sections, together with the stem, hog, transom and any bulkheads which are to be built into the boat. With single chine there will be one chine stringer fitted longitudinally on each side (marking the change in direction of the side panels); with double chine, two. The chines, bulkheads, etc., are all faired off so that the plywood panels will fit down flat and then the panels shaped and fitted with the use of glue and nails. The bottom panel goes on first, followed by the chine panel, for double chine, and then the side panel.

The other method, known as 'stitch and tape', works rather like a paper hat with panels joined along their edges being 'opened up' to form the shape. The panels of plywood (which must be shaped to small tolerances of accuracy for upon this depends the dimensions of the boat) are fitted in pairs. First the two bottom panels are joined along their centre edges by a 'stitching' of copper

Chine construction

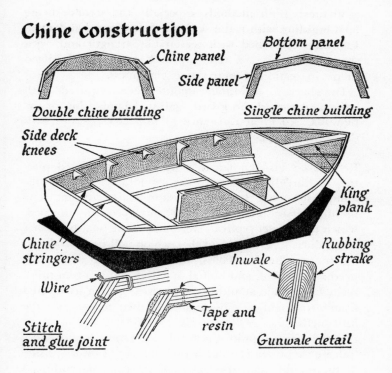

Chine panel

Bottom panel

Side panel

Double chine building

Single chine building

Side deck knees

King plank

Chine stringers

Rubbing strake

Inwale

Wire

Stitch and glue joint

Tape and resin

Gunwale detail

wire and then 'opened out'. The curvature of those edges results in a boat shape. Other panels are then added in pairs by the same method along with the transom and bow transom, or the sides are drawn together to form the stem. With the assembly 'lined up and braced', inch-wide glassfibre tape is then applied to the seams with polyester resin, first on the inside and then on the outside—a process which integrates the whole structure and makes it rigid. Further bulkheads, seats and buoyancy chambers are then added by the simple method of shaping the panels and fitting them in with more glass tape and resin.

Virtues: Both methods, especially the second, bring boat-building within the skills of the average handyman. Lightness combined with reasonable strength and, with the double chine, a fair approach to a round bilged shape in section. Comparative cheapness.

Drawbacks: Assiduous maintenance required with paint and varnish, to guard against delamination of the plywood. No great resistance to impact especially with the thinner plys.

Resinglass: also known as GRP (glass reinforced plastics) or fibreglass and, with the available car repair kits, almost too well known to require description. The builder's first job is to prepare a 'plug', probably in wood, which is an exact replica externally, of the boat he wishes to produce. The surface of this, which will be reproduced on the production boats, must be coated with polyurethane, sanded, and buffed to a high finish. From this a resinglass female mould is taken which is then set up in a frame, for rigidity. This mould is used to produce the hull shells.

The laying-up process is as follows. First a coat of waxy 'release agent' (to prevent it sticking) is put on the inside of the mould, then the 'gel' coat (special, un-reinforced resin), which forms the waterproof outer skin is applied by brush or spray. Fine glass-woven cloth is dabbed on with resin and then comes the bulk of the fibreglass matting, also resin-saturated. Having attained the desired thickness the inner surface is finished off with a layer of fine woven glass cloth and a 'gel' coat.

Other additives to the lay-up are 'fillers' in the mat to give 'body' and pigment, which is added just beneath the gel resin to produce the smart colouring. It is important to remember that only the outer gel coats are truly waterproof so that when the boat is in use extreme care must be taken not to damage these surfaces too much, or water will gain entry into the 'slightly porous' inner lay-up.

Virtues: There are no restrictions placed by the material (as with plywood) on the hull shape. Round bilged, imitation clinker, and versions of chine boats are adaptable to the material. Maintenance is minimal, but the surface must be treated with care. Good, replaceable keel and bilge runners are helpful, as is a wax polishing at the start of the season. Repairs can be made fairly easily with a 'kit'.

Drawbacks: Coldness to the touch. A rather rough interior (but see 'Double-skinned' GRP). Flexibility over large, unstressed areas and a poor resistance to impact where it is stressed. This is partly to do with the density of the material and the attempts to 'build down' to a weight. The more heavily built boats do not suffer in this way.

Double-skinned GRP: Many single-skinned resinglass boats have mouldings for a foredeck, side-decks, seats and buoyancy chambers which, when fitted, help to stiffen the hull and improve the interior appearance, as it is the 'mould' side that shows. An extension from this is when the whole interior, including a false floor, is made as a second moulding and bonded to the hull around the gunwale. The air space between the skins becomes a buoyancy chamber, to overcome one of resinglass's drawbacks— negative buoyancy. If the floor level is higher than the boat's waterline, then a port in the stern will allow any water to flow out and the boat will be 'self-draining'. If the space between the skins is filled with polyurethane foam, making a 'sandwich' construction, which bonds to both skins, then the whole structure is a homogeneous shell, with much greater stiffness and, as the stresses are spread over a large area, with a much greater resistance to impact damage.

Virtues: The same as for any resinglass boat but with much improved appearance.

Drawbacks: The amount of room inside is curtailed. A 'double bottom' (raised floor) can cause a very 'knees up' sitting attitude. Unless an 'air buoyancy' boat has her chambers compartmented, a hole anywhere in the outer skin will result in a total loss of buoyancy.

ABS: Hulls are thermo-formed into moulds from single sheets of the plastics material, ABS, by a vacuum process. Separate shells are made for the inner and outer halves and then sealed together, enclosing a synthetic foam filling. The restrictions of the process bias the manufacturer's choice of hull shape towards one which is well radiused and shallow with a rounded bow—not the ideal for sailing purposes.

Virtues: Lightness and toughness. Local damage can be repaired with a manufacturer's repair kit. No regular maintenance required.

Drawbacks: A lack of choice in the present market. As with resinglass, an owner would find some difficulty in fixing additional fittings.

Polyethylene: Similar in appearance to the kitchen-utensil material but of a high density, making it tougher. The hull is thermo-formed from sheets in a similar way to ABS. Skins can be bonded together to form enclosed air buoyancy, or it may be given a foam lining. The material has some buoyancy of its own. The wear, corrosion and solvent resistances are good. Repairs can be made by heat fusion, adding pieces of the same material, if necessary.

EPS (*expanded polystyrene*): Moulded entirely in high-density foam, the thickness of about 2 inches gives the hull its stiffness while the inherent lightness of the material gives tremendous buoyancy. The foam is of the 'closed cell' type so that damage, even to the point of being stove in, causes no absorption and destroys none of the buoyancy. In spite of having a toughened outer surface, which can be further protected by a coating of plastics, such as Isoclad, this 'soft to the touch' material is prone to damage. This can be repaired but appearance suffers.

Virtues: Lightness, a high safety factor and low cost. Because of the latter this almost becomes a 'disposable' boat.

Drawbacks: Appearance and the difficulty of keeping clean. The accidental spillage of some volatile liquids, such as petrol and methylated spirits, will soften or even dissolve the material. A short life and rapid depreciation in value.

Aluminium: There are few dinghies on the market in this material mainly because it is difficult to work and poses welding problems. The cost of the product then cannot compete with boats made in other materials. Nevertheless it has its advantages and probable advances in technology and methods may well make this a material of the future.

Virtues: Lightness, and imperviousness to rot or deterioration.

Drawbacks: Efflorescence, but this is not harmful and, if appearance is considered important, is prevented by painting. Anodising of the whole boat could be the answer but the boat is already expensive. It would take a lot of impact to cause any fracture but quite a small one could cause denting—a further debit to appearance.

With negative buoyancy some sort of buoyancy units must be strapped into the boat.

Inflatable rubber: These boats are at their best as yacht tenders, liferafts or being driven by outboard motors. There are sailing inflatables but their lack of lateral 'grip' on the water, together with a lot of 'wetted area' invariably makes them poor performers.

Virtues: Lightness and portability with compact storage at home. A good measure of safety but, as with any boat, not a factor to be abused by, for example, her occupants sailing her from a beach with a strong off-shore wind blowing.

Drawbacks: Performance, manoeuvrability and 'bouncy' motion.

HULL SHAPE AND PURPOSE

As already stated, materials can affect the shape of a hull, but within those restrictions a hull can be long, short, fat, narrow, deep or shallow, according to its weight-carrying and speed requirements. Of course, any boat can be put to any use, and there are some good 'all rounders', but it is as well to know to which extreme type the boat you are interested in is orientated.

Displacement and Planing Hulls: When a boat floats she displaces a volume of water equivalent to her weight and when she moves, that volume is continuously transferred from front to rear in the form of a wave. At any given moment that wave has two peaks, the bow wave and another at the stern and the distance between is the wave length. The speed of any wave is strictly related to its length and for the technically minded this is $1.35 \times \sqrt{\text{wave length}}$ (feet) with the answer in knots. So, since the boat-formed wave has a speed and the displacement hull cannot outrun that wave this also decides the maximum speed of the boat. The longer the boat the longer the wave and the faster she can go. This maximum is known as a boat's 'displacement speed'.

A planing hull, on the other hand, in certain conditions can go faster than her displacement speed by

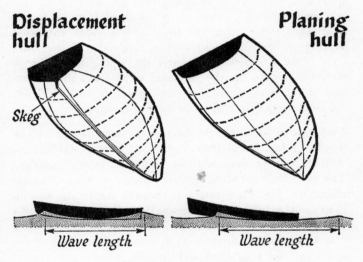

'skimming' on the surface. For this she must be light, have plenty of sail area and the wind to drive it and have plenty of flat underwater sections aft on which to skim. She still makes waves but dynamic lift means that these are less than her displacement waves and longer, since she can climb on to her forward part of the wave and leave the stern peak of it far behind.

The Beam Factor: If waterline length is important in deciding a boat's speed, her waterline beam is equally so.

The greater the beam, the more drag and the less speed. But the greater beam becomes useful by increasing weight-carrying ability and enabling the boat to carry more sail without being knocked over sideways. It also helps general stability.

Freeboard: The side of the hull above the waterline has two functions; the first is the obvious one of preventing spray and waves from slopping into the boat (and there is a great deal of comfort to be obtained from an ample freeboard); the second is to give the boat increased heeling resistance. As she heels, especially if the sides have 'flare' (angled away from the upright) the beam waterline creeps up the side and away from the centre-line of the boat, to increase the waterline half-width on the depressed side and resistance to heeling.

Underwater Sections: At the two extremes, these will be 'veed', giving deep draft and poor lateral stability or 'flattish', giving shallow draft and good lateral stability. The veed hull could be used for racing on inland waters where her low form resistance would accelerate well in light and fitful winds; the shallow hull would prefer stronger winds and planing conditions. Either or both may be tapered to finely veed bow sections to dig into the water and aid windward performance.

Most hulls compromise with something like a shallow 'D' in which the 'centre of buoyancy' moves in the direction the boat is heeled, supports the depressed side and so contributes to stability.

Chine boats approximate to round bilged sections and follow the same stability, speed and weight-carrying characteristics. Particular differences can be noticed in seaworthiness (they tend to slam into waves and ride less easily) and the resistance to heeling—which is more erratic through angles of heel.

Sections

*Veed hull – narrow
waterline beam –
low wetted area –
heels quickly*

*Flattie hull – broad
waterline beam –
stable initially –
large wetted area –
planes well*

*Shallow "D" hull,
flared topsides –
stable over a range
of heeling angles*

Underwater profile: If this is long and straight the hull will probably have good directional stability, that is, will need less rudder control to keep a straight course when sailing and will swing about less when being rowed or paddled. This is not all to the good and a keel may be more curved, descriptively called 'rocker', which will enable her to turn or tack more quickly and will ride more easily in a seaway.

Stem and Stern: In addition to the sectional fullness, or the lack of it, already mentioned, the ends may have 'rake' in profile. Vertical ends make full use of overall length for the best displacement speed but a raked stem makes for easier, and more seaworthy, bow sections and a raked stern may get a transom-hung rudder more under the boat, for greater manoeuvrability. For an outboard motor the vertical transom poses less problems.

Ends may be either pointed or blunt. The transom (or pram) bow (made famous by the Mirror dinghy) helps to keep the bows of a short dinghy buoyant and to give her occupants more space. Scows like the Fireball use a flat, shallow bow to help them skim over the water. The familiar wide stern transom and aft sections provide buoyancy, stability and a sharp break-away point for the water streaming under the hull of a planing boat. A displacement hull may be given a pointed stern which qualifies her for the name 'double ender'. She is not intended to go backwards any more than the normal dinghy, although she can show a disconcerting turn of speed in that direction to the unwary! The virtue is one of minimum drag so that she takes little wind to drive her, creates little wake disturbance and directs water on to her rudder to keep it well covered and effective. She loses on lateral stability though.

Keel and Bilge Runners: Although no more than an apology for the keels of her big sisters the dinghy keel takes the downward thrust of the mast and the stresses of the hull, houses and takes the stresses of the centreboard and case, and takes the main knocks and rubs of grounding.

Given a depth of 2 or 3 inches, it can also contribute to directional stability. This is accentuated if the keel terminates at the stern in a 'skeg'. Bilge runners also lend some directional aid but are primarily for the protection of the hull against scuffing on land.

DECKING AND LAYOUT

A dinghy without any decking is the true open boat and there is one famous class boat, the International Fourteen, in which this is a strict rule. But there are plenty of

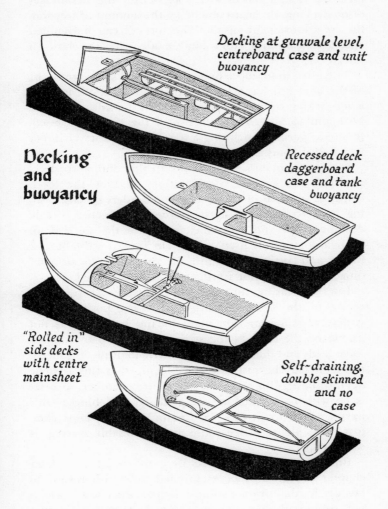

Decking at gunwale level, centreboard case and unit buoyancy

Recessed deck daggerboard case and tank buoyancy

Decking and buoyancy

"Rolled in" side decks with centre mainsheet

Self-draining, double skinned and no case

local one-design classes which also enjoy the advantages of no decking, the main one being the amount of freedom for crew movement inside. The foresail can be handed and access to and from land made over the bow in safety.

The filled in foredeck on many dinghies keeps spray out of the boat but has a useful structural function, making a very strong triangular 'box' to tie in the sides of a light boat and to absorb the stresses of the rig. A 'recessed' foredeck, perhaps enclosing a forward buoyancy chamber is a useful compromise.

Similar comments could apply to side and stern decking with the further considerations of how the level, width or absence of these, helps or hinders special boating interests. For rowing or fishing the absence of side decking helps and the same applies to the stern if an outboard motor is used much. Decking is valuable for racing and 'sitting out' while, if cruising is the primary object, lower, interior seating would be preferred.

Interior seating includes thwarts which traverse the centre of the boat and provide rowing positions and structural stiffening. On many racing boats this is not usable by the crew as the space is taken up by fittings, sheet leads and the horse for the centre mainsheet. Indeed, the crew may be justified in believing that this is some devilish design for a mediaeval torture chamber, so beset is he with boom, kicking strap, sheets and blocks with nary a perch to sit on! To forestall mutiny, crew comfort, with plenty of leg room and reasonable seating, is worthwhile.

Buoyancy will be covered in more detail in a later chapter but it must be mentioned now with regard to layout. By definition—volume lighter than water which will support the boat when swamped—it must take up a fair amount of space. If this serves a dual purpose such as side or rear seating, provides structural strength, or will bear judicious use as stowage so much the better. This goes only for built-in buoyancy in the form of rigid chambers of course; the plastics bag buoyancy is a strictly one-purpose commodity.

Many racing dinghies have what is known as 'rolled in' side decks in which the decking and inner panel form a buoyancy chamber each side. For lightness, strength and crew comfort the change in plane from deck to panel is given a soft curve. It presents a clean interior layout but for cruising and carrying extra persons one misses the additional seating provided by side deck and benches.

The floor space must be thought of as part of the layout, for although dinghy sailing is done mostly in the sitting position there is an amount of moving around which calls for adequate foot room and non-skid surfaces. A level floor is a great comfort on a cruising boat or one used for fishing or outboard motoring. If a boat is large enough to be used for camping and sleeping aboard the same applies and the addition of floorboards, slatted or solid sheet, is one way to achieve flatness.

The centreboard case occupies a dominating position in the boat. There are strong leverage strains during sailing and it needs to be well braced by a thwart or two, or knees against the floor fore-and-aft. Its space requirements make its substitution by a daggerboard and case (lifting straight up and down instead of hingeing) a thought for a boat not required mostly for racing, but one has instead the encumbrance of the board lying around in the boat when not actually sailing.

THE RIG

The sail plan is usually designed with the class of boat but in some cases a choice is available, say between gunter and Bermudian on a general-purpose boat. In recent years the adoption of Bermudian has been universal on new racing-dinghy designs and many older craft with other rigs have changed too. The efficiency of this rig, particularly to windward is undoubted and most research on sail and mast control is done on it.

Nevertheless there is an element of fashion in this which is understandable but illogical. Since racing is competition between boats similarly rigged the higher potential speed of the more efficient rig will not alter the standards between crews, nor increase their enjoyment.

A boat not being used for racing may still be influenced by fashion, through its designer or owner, but at least the choice is there. Considerations affecting that choice will be efficiency, ease of rigging, number of crew to work sails, ability to reef or reduce sail, mast height for a boat left on moorings, appearance, ability to stow spars inside the boat and many others.

Bermudian Sloop: Characterised by the single tall mast supporting the triangular mainsail and the gooseneck for the boom which spreads the foot of the sail. The luff of the sail is usually attached by a mast-track or a groove and the foot similarly although it may be attached only at the ends (tack and clew) and be loose-footed. There is one foresail, without a boom, attached at the bow and a point on the mast. This point varies between dinghies from 'two-thirds up the mast' to the very top (the masthead rig). This foresail is commonly called a jib. If a second foresail (or true jib) is set on a bowsprit this becomes a *cutter rig*.

Virtues: Simplicity and efficiency both for rigging and sailing.

Drawbacks: The mast, which causes windage problems on the moorings and in the dinghy park and its transport

Rigs

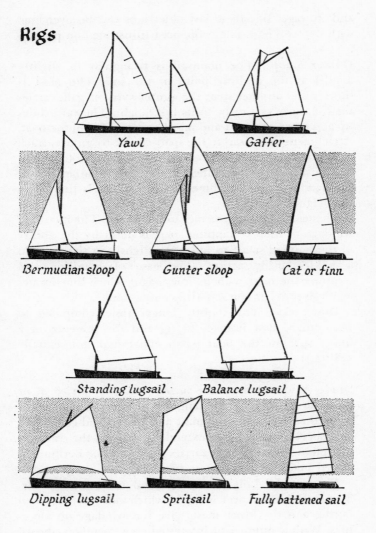

Yawl

Gaffer

Bermudian sloop

Gunter sloop

Cat or finn

Standing lugsail

Balance lugsail

Dipping lugsail

Spritsail

Fully battened sail

and stowage. But these last objections can be overcome with the 'two part' telescopic mast (illustrated on p. 104).

Gunter Sloop: The mainsail is triangular or slightly angled at the midway point of the luff. The mast is shorter and another spar, the gunter yard, or gaff, carries the luff, near vertically, to its full height. The yard slides up and down the mast and may be carried flat in the boat. Attachment of the sail to the spars can be by groove, track or by lacing. The foot goes on a boom, loose footed or by groove. The foresail differs from the Bermudian only in the height of the attachment point, which is limited to the height of the mast.

Virtues: A good performer to windward. The increased windage of the overlapping spars is offset by the ability of the 'jointed' mast to fall away slightly in the gusts in a self-feathering action. The ease of handling and stepping the mast with its low height which ameliorates windage problems on moorings and ashore.

Drawbacks: The slightly longer time, compared to Bermudian, that it takes to rig and the presence of a bulky spar in the boat when afloat and not actually sailing.

Yawl: A second mast, the mizzen mast, is carried at or near the transom. The triangular mizzen sail is laced or tracked at the luff to the mast and spread and controlled by a sheet to a fixed spar extending aft from the transom, a bumkin. The foremast carries the gunter or Bermudian rigged mainsail and one or two foresails.

Virtues: All sails are smaller and more manageable. The spread of sail area is lower, which aids stability when sailing and the lower masts give less windage on moorings. With a number of interested crew members aboard there are more sail handling jobs. A reduction of sail

46

area is instantly achieved by using only one foresail and mizzen or some other combination.

Drawbacks: Suitable only for the larger dinghy and when maximum sail efficiency is not the main criterion.

Ketch: Similar to the above and, once again, a suitable rig only for the larger dinghy. The distinguishing point is that the mizzen mast is set further forward of the transom (technically forward of the tiller post) so that the bumkin may be unnecessary, as the sheet for the mizzen sail goes to the transom.

SINGLE-SAIL RIGS

Mostly for the single-handed sailor either in a boat designed for one, or in a larger boat with non-working passengers.

Cat, Una or Bermudian Sail: The popular rig for solo racing dinghies which combines simplicity with efficiency. This is basically the Bermudian sloop rig without the foresail.

Virtues: Very good windward performance which is thanks to a long straight vertical luff behind which most of the sail drive is developed.

Drawbacks: The height of the rig which reduces stability when sailing plus the disadvantages of the Bermudian mast ashore. Downwind the sail area is all on one side, whereas two sails can be 'goosewinged' either side of the mast, with less tendency to broach to—which occurs when the wind takes over control of the boat from the rudder and swings her into wind, with the possible consequences of the boom dipping into the water and a capsize.

Balance Lugsail: The sail is four sided, the shorter two

sides being 'leading edges'. The upper of these, the head, is attached to a yard which is hoisted and projects about a third forward of the mast. The luff (the lower of the two) is kept taut by the downhaul of the boom, to which the foot is bent. This spar also has part of its length forward of the mast.

Virtues: In use, the sail reduces the pull on the sheet by having some of its area forward of the mast so that wind pressure equalises itself to some extent. This pressure forward of the mast also helps to prevent the peak of the sail falling off to leeward, as happens with a gaff.

Drawbacks: The less efficient windward performance due to the backward inclined luff and that on one tack the sail is pressed against the mast giving a disturbed flow across it for the wind.

OTHER RIGS

Standing Lugsail: Like the balance lugsail, but the boom does not go forward of the mast, the luff being inclined backwards from the throat to the tack.

Dipping Lugsail: Rather like the balance lugsail with the difference that the yard is carried further forward, the tack of the sail is hooked to the windward gunwale and it is boomless. When changing tacks the fore part of the yard must be 'dipped' behind the mast so that the mast is always to windward. An awkward procedure but the lifting performance of this sail on a reach is equalled by no other.

Gaffer, Gunter Lugsail or Lugger: Has a yard which supports the head of the four-sided sail at an angle (about 30 to 50 degrees) from the mast. There is a boom and no part of the sail is forward of the mast. A foresail is usually used. It is simple and used much by small knock-about dinghies in which its windward performance is probably better than that of the balanced lugsail. Fitted to larger dinghies, the space above the yard can be filled, in light weather, by a triangular topsail.

Spritsail: Gets its name from the sprit, or yard, which hinges from low on the mast to the peak of the four-sided sail. It is usually boomless. The part of the sail above the yard is hoisted by a halyard sheaved to the top of the mast. A working fisherman's sail, he can keep the cockpit clear yet render the sail temporarily inactive by 'scandalizing' it—hoisting the sheet to the sprit or pulling the sprit vertical. A tiny dinghy like the Optimist uses it to get the maximum spread for its sail from a very short mast.

Fully Battened Sails: Most modern mainsails have short battens (wood, plastics or carbon fibre) in pockets at right angles to the after edge, or leech. This is to support the outward curve of the 'roach' and to prevent any inward curling of the cloth. But some sails are 'fully battened' from luff to leech, notably those on catamarans, to pro-duce a stiffer sail and one that can have its curvature controlled by tension applied by a cord between the sail and a projecting portion of batten at the leech.

THE MAST AS PART OF THE RIG

On lugsail-rigged dinghies and most of the older types the mast is merely a sail support. On the modern Ber-mudian-rigged dinghy it has become a factor in sail con-trol and shape (its curvature, in section, from luff to leech).

The unstayed mast has no shrouds to guy it and is held

in a rigid anchorage (yet may be allowed to rotate in it) between keel and deck level. Above that it relies on its own strength to withstand wind pressure. A knock-about dinghy may adopt the method for simplicity but for a racing dinghy the advantages are the lack of wire windage, the streamlining effects of a rotating mast and the bend adopted by the mast due to its taper and the pressures put on it by the sail. This bend flattens the sail for windward sailing and allows it to deepen again downwind, when pressures are released and it straightens.

The stayed mast has a forestay and a shroud each side, while the largest dinghies may even have a backstay as well. These support the mast so that it may be stepped on the deck instead of the keel. In addition the mast may be given further staying by spreaders, rods between mast and shrouds, and 'diamond' staying, a wire on each side, attached to the mast at the top and fairly low down and strutted with rods. Very generally, the spreaders support or control the lower mast bend while the diamonds do the same for the upper mast.

This staying and the controls on it can induce various amounts of curvature, or flow, into the sail and when combined with a fully battened sail is known as a 'hard rig'. The unstayed mast, in contrast, when the wind it-self does much of the shaping and the sail suffers less mechanical constraints, is known as a 'soft rig'.

3

The Choice of a Dinghy

With the objective review of dinghies behind us we can get down to specific boats and individual choice. To do this rationally means examining one's own needs, the storage available, the water one is to sail on and one's finances, after which the choice should be easy. But I know, being human, that choice seldom works like that. Indeed, given those eliminating tests, I should not be sailing the dinghy I do!

Instead, choice is more often a matter of whim, availability, the class sailed in the club, or the influence of a friend or a salesman, and, apart from the last one, I would not want it otherwise. But before finally opting, it is always as well to sail in the proposed boat, so that really dreadful mistakes are avoided and you or your family are not put off sailing for life.

TRIAL SAILS

Most builders will arrange for a demonstration of their boat and if it turns out to be in a sheltered backwater on a very calm day, ask for a repeat trial in better circumstances on a better day, rather than accept the salesman's assurances. Or try another builder.

Representatives of class associations, which administer the class rules, measurements and fixtures, can often direct prospective recruits for the class to members willing to show off their boats. There are none so avid in promoting class boats as association members, as design sponsors know to their benefit!

A GENERAL IDEA OF WHAT YOU WANT

After a few trial sails, discussion with members of the trade and sailing acquaintances, with observation of the sailing scene from press reports and sailing centres and the suitability of the sailing area you intend to use, an idea of a suitable craft should begin to form. Apart from monetary considerations, which will be discussed later, the guiding factor will be the use to which the boat is to be put together with the ability of you and your crew to fulfil that use.

Racing: If this is your aim, but you still have to learn to sail and lack some of the resilience of youth to off-putting 'disasters', it would be rather unwise to go for an out-and-out racing dinghy like a 5-0-5 or a Hornet right away. One *can* learn in these boats, but it would be better to start with something a little less potent like a GP 14 or an Enterprise until one gains competence in handling and in sailing in the close company of other boats.

Racing dinghies in this country are not usually crewed by more than two (multi-crews are the domain of the keelboat classes, outside the scope of this book), but the Jollyboat and the Wayfarer are boats which could take three or four and still race, although it should be remembered that the weight handicap would put such boats at a disadvantage against the two-handed boats in the fleet. Again, if you are inexperienced it would be better to start with a smaller boat and then graduate.

Starting single-handed racing is usually less of a problem as one can more easily accept responsibility for oneself than for others. Yet care is still required because the inability to control a 'difficult' boat could mean putting others at risk in a rescue attempt. Single-handers are mostly light and responsive but some are more so than others. The Solo is a well-tried boat without being too extreme, while in the surfboard department the Minisail, the Beachcomber and similar boats are not too hard for the beginner. The more experienced sailor will consider a boat with more sophisticated equipment and speed potential. Among these are the International Finn, the International Moth and the International Contender which may have sitting-out aids like sliding seats and trapezes.

Multihulls, chiefly catamarans, now hold an important place in the racing scene and if you are attracted by sheer exhilarating speed over water and have strong nerves and rapid reflexes, one of these will be your ultimate objective. What you must accept is some discomfort from spray (the inevitable concomitant of speed), and rather horizontal sitting and movement about the boat, the top of which is inevitably flat. Their speed, coupled with their slowness of tacking make them of doubtful value for learning, but their high initial stability is a point in their favour.

There are some 'tame' cats which could be considered as multihull initiating boats. Among them is the Watercat, built in marine plywood in 10 ft and 12 ft versions and Felix, a 14-footer with resinglass hulls and tubular alloy beams to bridge them together. But neither is regularly raced yet, to my knowledge.

Going on to the serious racing classes, the most popular catamaran is the Prout Shearwater with her smaller sister the Swift, a long way behind but still second. The Swift makes a good single-hander if crews

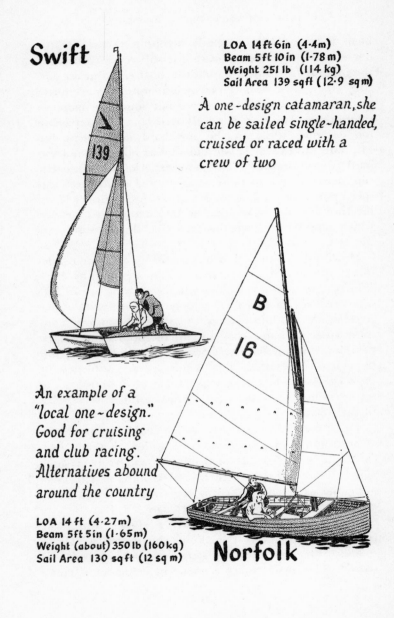

Swift

LOA 14 ft 6 in (4·4 m)
Beam 5 ft 10 in (1·78 m)
Weight 251 lb (114 kg)
Sail Area 139 sq ft (12·9 sq m)

A one-design catamaran, she can be sailed single-handed, cruised or raced with a crew of two

139

An example of a "local one-design." Good for cruising and club racing. Alternatives abound around the country

LOA 14 ft (4·27 m)
Beam 5 ft 5 in (1·65 m)
Weight (about) 350 lb (160 kg)
Sail Area 130 sq ft (12 sq m)

B 16

Norfolk

are in short supply. More sophisticated is the Unicorn for which meetings are held, mostly around the South Coast and the Thames Estuary.

There is one useful trimaran to note, the Anderson Shark. This is a 14-footer in moulded and marine plywood and each of the float supporting struts, either side of the main hull, is used in turn as the sliding seat support. The Shark is a single-hander, and although her special sailing characteristics may take some getting used to, there should be little to dismay the beginner and plenty to satisfy the enthusiast.

It should be appreciated that catamarans do need a fair amount of sailing water and are thus unsuitable for narrow rivers. Only on open water can their speed be developed and their lack of manoeuvrability be accommodated.

Pottering and Cruising: If you do not want to race, and feel that boat-speed should not be gained at the sacrifice of stability, then you should look for a dinghy with a generous beam for her length, firm sections and a low aspect-ratio sail plan of modest area, or one which can be reefed without trouble.

Yet there are no rules and as most boats can be raced, so most dinghies can be cruised.

At one end of the spectrum is the heavily built cruising dinghy. It may be your ambition to emulate the exploits of the cruising dinghymen in this country, who sail boats across the Channel and the North Sea or just to do a little 'coast hopping' from port to port in fine weather. Well, when you are adequately prepared and have acquired some experience you can, but there is nothing wrong in getting the right kind of boat right away.

The Wayfarer is one such boat and the open-sea voyages

of Frank Dye are enough proof of the seaworthiness of this class. The hull is double chine and built in marine plywood or resinglass. She is just under 16 foot and has the weight to carry through waves. Traditional clinker building in wood is immensely strong for this kind of dinghy but, unfortunately, due to the man hours and skill required for building them, they are getting more and more expensive and rare. Nevertheless there are small builders still prepared to turn out the 'one off', often more for love than profit and one must just hunt them out. L. H. Walker & Co. of Leigh-on-Sea, Essex, is one firm which turns out a range of clinker dinghies of which the 12- and 14-footers are deservedly popular.

Lapstrake is a form of clinker but with fewer and wider planks each side. The fishing 'coble' is built in this way and makes a fine boat for beach and coastal sailing. Then there are the Drascombe boats built by Honor Marine. These were originally built in wood but are now also produced in resinglass whilst retaining the lapstrake appearance. Largest is the Longboat, which at 21 ft 9 in and with a cuddy takes her well into the yacht category. The Lugger at 18 ft 9 in is a little more dinghy-like in proportions, while the Dabber at 15 ft 6 in is more compact. All are good sea boats.

Simulated clinker is resinglass moulded to give a wooden clinker appearance and by so doing the owner gets the aesthetic appeal and the spray damping effect of the clinker form coupled with the cleanliness, the ease of maintenance, and the resistance to decay and marine borers of the resinglass. Yet nothing is perfect in this world and I believe that some strength loss, compared with wood, is the fly in the ointment. I have yet to be convinced that mere wrinkles in the lay-up give as much strength to the hull as overlapping planks, with or without riveted fastenings. However, compared with

Wayfarer

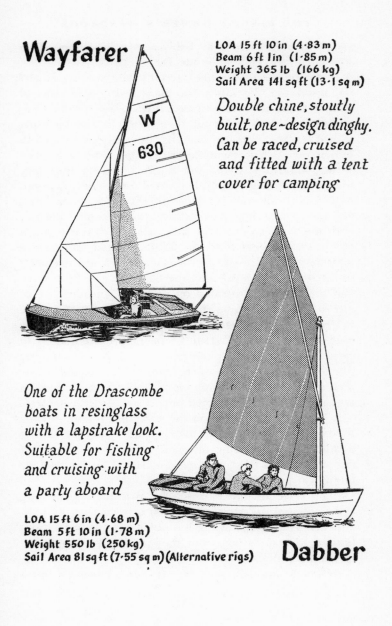

LOA 15 ft 10 in (4·83 m)
Beam 6 ft 1 in (1·85 m)
Weight 365 lb (166 kg)
Sail Area 141 sq ft (13·1 sq m)

Double chine, stoutly built, one-design dinghy. Can be raced, cruised and fitted with a tent cover for camping

One of the Drascombe boats in resinglass with a lapstrake look. Suitable for fishing and cruising with a party aboard

LOA 15 ft 6 in (4·68 m)
Beam 5 ft 10 in (1·78 m)
Weight 550 lb (250 kg)
Sail Area 81 sq ft (7·55 sq m) (Alternative rigs)

Dabber

a smooth hull, there is probably a gain in stiffness. One simulated clinker boat is the Devon Yawl.

It may be of only passing interest to the cruising dinghy sailorman but the lands and increased wetted area of clinker boats do make them a little slower than their smooth-hulled sisters, a fact evidenced by the Yachting World Dayboat. She can be built either clinker or smooth hulled, and when racing together the latter are given a time handicap. Given the choice of which hull type to copy, the resinglass builders opted for the faster smooth hulled boat, but all versions are magnificent sea boats and have the weight and stability to lie to moorings.

If you wish to camp or sleep aboard, while lying at moorings, the boat needs to be flat floored and have a tent cover. Most of the already mentioned cruising boats could be so equipped and the Seafarer, the Mirror 16 and the GP 14 (illustrated on page 14) are others. This sort of boat needs to be at least 14 ft long and even then, except on placid inland waterways, the motion will be enough to rule out restful sleep. The alternative is to carry a small tent and pitch on dry land. A true camping dinghy must also have such features as stowage space, a shielded area for setting up a stove, a small paraffin pressure or methylated spirit burner and other 'comforts'.

A heavy dinghy is fine on the water but is troublesome if you have to trail and launch her on every sailing occasion. A contrasting alternative is a boat on a roof rack but only a light boat can be carried on the average family car.

Light boats include the Bobbin in resinglass, 9ft long and 95 lb and the Mirror 10 which is in marine plywood and weighs 98 lb. Cruising has obviously got to be more limited with these boats but they are handy enough for forays from the beach.

As weight and size goes up a trailer is necessary but a

Gull

LOA 11ft (3·35m)
Beam 4ft 9in (1·45m)
Weight 172lb (78kg)
Sail Area 70sq ft
(6·5 sqm)

*Double chine one-design.
A brave little boat for
pottering, especially
suitable for children*

V
1549

*One-design, surf-
board type. Easy
to carry, rig and
sail. Mostly for
single-handed*

1731

LOA 11ft (3·35m)
Beam 3ft 10in (1·20m)
Weight 84lb (38kg)
Sail Area 56sq ft (5·2 sqm)

Topper

host of very useful two- and three-man boats now comes within consideration. Among these is the Gull, in plywood or resinglass an 11-footer weighing 170 lb, the Heron, also with the material choice and a similar length weighing 150 lb, and the Skipper 12 and 14 dinghies in resinglass weighing 145 and 160 lb respectively.

For a single-hander there is a choice between a reduced-sail version of one of the aforementioned dinghies and a sailing surfboard. These little boats have already had a mention with regard to racing but they are also good for launching off the beach and for doing some cautious pottering around tidal channels, where their shallow draft and lightness for portage helps. The Topper and the Sunfish are boats in this class.

Rowing and Outboard Motoring: Most racing dinghies have little or no alternative power to sails. A paddle is the usual recourse but is fairly inefficient even when two are carried and used. For crowded anchorages, for getting off a leeshore and for general all-round experience in boatmanship there are nothing like oars, which means fitting rowlock sockets. There are boats which row well: boats with the directional stability provided by a straight keel, a skeg and bilge runners. A boat with these qualities is the Small Craft Shetlander, a resinglass 14-footer which also sails and motors well. In short, a good all-rounder.

An outboard motor pad can be fitted to most dinghies and for the low powered motor, say, 2 to 3hp (enough to get a dinghy out of an anchorage or back from a fishing trip when the wind has died) a dinghy transom does not need much in the way of padding to strengthen it. More of a problem is the stowage of the motor when not in use. (See Chapter 7.)

Larger dinghies, such as the Drascombe range of boats and the Sussex Cob, have an outboard trunk, an open-bottomed box forward of the transom into which the motor is lowered or tilted, but the arrangement requires space.

Fishing: If you plan to do a good bit of sea fishing from a moored boat, some useful attributes to look for are very full sections, weight, a flattish floor or floorboards, and a daggerboard in preference to a hingeing board, as the case takes up less room. The ESB Naiad, a 13 ft resinglass dinghy which also has a double floor and is self-draining from a swamped condition, certainly fulfils these conditions while another is the Foreland 12½-footer.

ONE-DESIGN AND RESTRICTED CLASSES

These two terms have very distinct meanings:

One-design class boats are intended to be identical in every way. But since boats are more complicated than buttons this is rarely possible. Resinglass hulls get nearest as they are produced from the same mould, but even then there may be differences in the lay-up which may affect performance. In wooden boats, depending on the water content of timber, the thickness of a sawcut and the skill of the builder, the differences can be more marked. So, for each class there is a list of measurements and permissible variations, if any. Included are hull dimensions, weight, buoyancy checks and sail and spar measurements. Each boat is checked by a measurer recognised by the RYA, and a certificate issued by that body if everything is in order. Some builders can arrange for this. The possession of a class certificate enhances the value of the boat and without it, she cannot be raced officially.

Restricted class dinghies are not intended to be identical. Certain limits are set, such as length, weight and beam of the hull and the area (but not the plan) of the sails, and design and development allowed within those limits. So you can get a Merlin Rocket (illustrated on page 15), for example, with a full or a narrow sectioned hull or a large foresail with a small mainsail or vice versa. The layout of the decking and interior may vary from others and the mainsheet may be to the transom or central. The centreboard and rudder may be individual to the boat, too.

Other restricted classes include the National 12 ft Class, the International 14 ft Class and the International Moth. Once again, these boats must be measured and be found to conform before being issued with certificates.

So when choosing a design from within these classes care must be taken that the type of water you intend to sail her on, the strength of wind that is usual, the weight of you and your crew and the agility available to control her are all compatible with the boat. For all these things were factors in the original design.

LOCAL ONE-DESIGNS

Every area of sailing water has its own characteristics. The nature and depth of the water; its exposure to wind from certain directions or its shelter, the topography of the shoreline with its 'channelling' or rebounding effects on wind; all these and more considerations influence the design of the local class boat, like the Norfolk illustrated on page 54.

Examples abound in this country and abroad, and there is no better boat to buy than the local one-design, if 'local' is where you intend to sail. It may still be suitable if the conditions are similar to the area which bred the design but to sail a boat designed for a sheltered river off a rugged, exposed, deep water coast is risking an unhappy boat, and owner.

LOA 12 ft (3·6 m)
Beam, min. 4 ft 6 in (1·37 m)
Weight 200 lb (91 kg)
Sail Area 90 sq ft
(8·4 sq m)

National 12

Restricted class racing dinghy. Ease of handling depends on the design, which is sometimes "customerised" for crew weight and sailing water

Restricted class single-hander. Many designs, some with "trampolenes" or "wings". Difficult to sail

LOA 11 ft (3·35 m)
Beam, max. 7 ft 4 in (2·24 m)
Sail Area 85 sq ft (7·9 sq m)

International Moth

PRICES—THE CRUNCH

Class Boats: The difference between what you would like and what you can have will probably be set by your pocket. If your first choice is a class boat which can have various builders (with a sole agent or builder you have no opportunity to 'shop around') get as many quotations as you can. Try to visit the yards and examine boats on the beach and in the dinghy park, noting the workmanship and builder's name-plates. Take full account of what each builder is offering as parts of the standard boat and what will be needed in the way of extras. Sails are seldom included, neither are oars or paddle nor any 'tiddly' fittings, such as self-bailers or jamming cleats.

Some builders do two versions at different prices, the standard boat with a wooden mast and a de-luxe boat with all the gear and an anodised alloy mast.

Some 'fancy' accessories can be added 'as you go', others will be tempting extravagances while a lot will be necessities. I have made a list of reminders (without prices for which you can insert the current figures) that could be for a simple boat without trapeze or centre mainsheet, something like a GP 14. Add up the current prices against your selection, plus something for fitting charges where required, and the total may well come out to an additional third to a half of the cost of the boat!

Sails (standard)
Genoa foresail
Spinnaker and fittings
Spinnaker pole or jibstick
Spinnaker chute or net
Polyurethane paint in lieu of conventional
Signwriting boat's name

Painting 'coach line' along sides
Brass or plastics keel and bilge strips
Non-skid floor strips
Painter
Anchor and warp
Kicking strap with jamming tackle
Toestraps
Jamming cleats with sliding fairleads for jib sheets
Full width sheet horse
Ratchet block for mainsheet
Self-bailers or transom flaps
Tensioning lever for main or jib halyard
Flag halyard and racing flag or burgee
Mooring cleat for foredeck
Rowlocks and sockets
Oars or paddle
Sponge and bailer
Fenders
Boat cover
Outboard motor pad
Outboard motor
Launching trolley
Trailer
Tying down straps
Lighting board and cable
Car number to board
Towing bracket for car
Ball hitch for towing bracket

There may also be a measurement fee and a designer's royalty.

If you elect to get the measuring done yourself, and there are club measurers empowered to do this, be sure that the boat is accepted from the builder 'subject to obtaining a Class certificate'. He will agree to this but may set a time limit, three months or less, so be sure to get it done quickly.

Traditional and One-Off Boats: If this is the type of boat for you then the simple solution is to visit the local boatyard and state your requirements. Often he will have plans, photographs or an example of the occasional one-off he does, or he may recommend another yard or a designer.

If you obtain the plans of a boat which takes your fancy (the yachting press will provide the names of agents who will send illustrated lists of plans) the task is to get a builder. Again it pays to get quotations, and you will no doubt be surprised by the variation in the estimates. Usually a designer's royalty fee is included in the cost of the plans and this is for the construction of one boat.

Finish Her Yourself: You may not think of yourself as much of a boatbuilder but finishing a boat should be in the compass of most handymen and the incentive is that you save money. Many builders will supply completed hulls for finishing by the purchaser and how much you save depends on how much there is to do. In the minimum case, where only the fittings have to be screwed on, the saving might be only about a tenth of the complete boat price. Many resinglass boats are offered in this condition. With a little more work, perhaps the sanding and varnishing of the woodwork including the centreboard and rudder blanks, the saving would be about a seventh. This could still be a resinglass boat or a nearly complete wooden one, either class or traditional clinker.

Big savings occur if the decks remain to be fitted, about one third of the purchase price. This could mean quite a lot of work but the materials are supplied and cut to shape. Composite boats (wooden decks on resinglass hulls) are offered this way, also many wooden class boats. As with complete boats, just what is included in the way of fittings and rigging must be carefully noted, for an 'apparently cheap' kit may be devoid of major items such as spars. Fittings can amount to a third of the price of a complete boat and a half to two thirds of a kit boat!

Building from Scratch: This is where the real savings come in, yet, unless you are also going to make some of the fittings and the sails, the most you can save is half the purchase price of the complete boat.

Construction methods have been noted in a previous chapter so I need only add that the easier the construction method the less you save. Build a 'stitch and tape' plywood boat from scratch and you will save about a fifth on a 'ready to sail' boat, whereas with a framed chine construction the saving would be about a half.

These savings are by using kits supplied by builders with parts and panels cut to shape and including fastenings, fittings and paint and varnish to complete. One can just buy the plans from the designer and get one's own materials but with the wastage that occurs from cutting 'one-off' requirements from rectangular sheets of ply, this does not necessarily work out cheaper. But the ability to buy as one builds can spread the cost.

Moulded plywood boats, built up of diagonal veneers, are good for the home constructor with a fair amount of time. Kits, other than completed shells, are not available, and as a building mould must be carefully made and then thrown away after use it would be economical to associate with friends so that the mould could be used to produce more than one boat. Otherwise the cost is higher than with chine or clinker building—which need only rudimentary frames.

A similar case could be made out for resinglass boats

or constructions with an EPS core which is then resin-glassed outside and then in. Incidentally, you cannot take a mould from an existing boat without infringing the designer's copyright. More information about building boats from resinglass can be obtained from the material suppliers.

THE SECOND-HAND BOAT

Most dinghies have three or four owners during their lifetime, so it follows that more boats are acquired second-hand than new. Transactions originate through the yachting press, boat auctions, boatyards (having been traded in for new) and friendly arrangements between acquaintances and sailing club members.

As already noted, boats hold their value better than many other possessions—in fact they have been doing rather better than the pound in recent years, giving the erroneous impression that they appreciate. Reference to new boat prices soon corrects that one!

Age, condition and the amount of ancillary equipment are the main factors governing price. Age is deceptive and a lot depends on the stoutness of the building and the care that has been taken of the boat. There are pre-war traditional dinghies in good condition which go cheaply and marine plywood chine boats only two years old that look almost worthless which are given a high asking price.

Deterioration in marine plywood boats is shown by the opening up of the surface grain and delamination. The hull and decks at the transom are vulnerable points. There may also be badly patched repairs to impact damage or scoring to the hull bottom. These things affect value and while the damage can be made good, even to the extent of renewing decks and panels, allowance should be made. The base of the centreboard case is where rot can develop.

Solid wood boats may be rotten in more places but unless there are built-in buoyancy tanks it should not be difficult to detect. The fingernail or the penknife tip will tell all. The covers or bungs of buoyancy chambers will exude mustiness if all is not well and the darkening of the timber sides near the floor and sides is another sign.

Comparisons and value estimates are made difficult by the amount and quality of fittings and extras. A boat with two sets of sails, all the latest 'gear' a trolley and road trailer may rightly have a price tag double that of a similar boat without that equipment. It is for you to decide how much more you would have to spend on the cheaper boat to bring her into sailing trim and compare the result. On the other hand it may be possible to bargain a lower price by not taking some of the superfluous equipment of the well-endowed boat, or even to buy the lot and sell off later what you do not need.

Resinglass boats, if they were originally well built, should deteriorate no faster than wooden boats but, due to loss of gloss and the crazing of the surface, they appear to, with a depreciating effect. Points to look for are the scoring of the bottom, from being dragged over rough ground (this should be prevented by wooden, replaceable runners) 'puttied' repairs (instead of the full lay-up treatment) and the cracking and 'coming apart' at joints, particularly at the base of the centreboard case and around buoyancy tanks.

CLASS MEASUREMENT CERTIFICATE
AND OWNERSHIP

The fact that a certificate enhances the value of a class

boat is worth repeating. If a dinghy has not been measured a price can be agreed on a 'no measure, no sale' basis. If she has failed measurement she may be cheap on that account and if you do not want to race this may be a means of acquiring a boat, but remember you will have the problem when your turn to sell comes round.

Another use for the certificate is that it is some proof of ownership. The thieving and selling of marine equipment and boats is always a problem and the second-hand boat-buyer must be on his guard. The builder's receipt is another proof but this seldom gets further than the first owner. It would be useful if legislation made it obligatory to have a log-book to go with each boat, as for cars. But with a surfeit of legislation in this country I suppose that would be a sad happening too.

4

Sails

Sails to harness the wind are one of the oldest sources of natural power known to man and on boats it was once quite a simple matter. A sheet was set up square to the wind which just blew the boat along. Unfortunately this only worked downwind or across the wind and even ships of the Middle Ages got embayed in shallow depressions of the coastline, instead of being able to tack their way out.

The fore-and-aft rig changed all that and boats were able to sail closer to the wind and tack their way to windward. So important is this point of sailing that most of the development in sail and rig in recent years has been directed towards it and the modern yacht and dinghy can sail to within 45 degrees of the true wind (see pages 193-4). For the racing man who must sail a proportion of every race to windward this is justified, for there is more time to be gained on the beat than will be lost on the run, but for the cruising man it may not be so important.

The Action of the Sail: A sail divides the wind passing over it and when inclined slightly to that flow causes pressure to build up on the exposed (windward) side and to decrease on the sheltered (leeward) side. This pressure differential, minus the drag of the spars and rigging, is

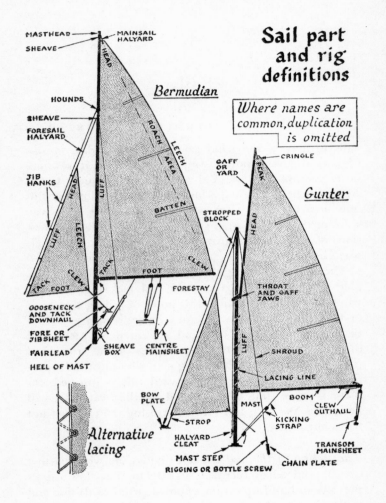

Sail part and rig definitions

Where names are common, duplication is omitted

MASTHEAD
SHEAVE
MAINSAIL HALYARD
HEAD

Bermudian

HOUNDS
SHEAVE
FORESAIL HALYARD
HEAD
LUFF
LUFF
ROACH AREA
LEECH
BATTEN

JIB HANKS
LUFF
LEECH
CLEW
TACK
FOOT
TACK
FOOT
CLEW

GOOSENECK AND TACK DOWNHAUL
FORE OR JIB SHEET
FAIRLEAD
HEEL OF MAST
SHEAVE BOX
CENTRE MAINSHEET
FORESTAY

STROPPED BLOCK

GAFF OR YARD
PEAK
CRINGLE

Gunter

HEAD
THROAT AND GAFF JAWS
LUFF
SHROUD
LACING LINE

BOW PLATE
STROP
MAST
BOOM
CLEW OUTHAUL
KICKING STRAP
HALYARD CLEAT
MAST STEP
RIGGING OR BOTTLE SCREW
CHAIN PLATE
TRANSOM MAINSHEET

Alternative lacing

the effort which drives the boat. This effort is roughly at right angles to the chord of the sail so that when the sail is sheeted away from the centreline of the boat, as on a reach, the effort is directed more forwards than sideways. When the sail is drawn in closer to the centreline of the boat on the other hand, as on a beat, the effort is directed more sideways than forwards.

The sideways effort is counteracted or absorbed by the lateral resistance of the hull and centreboard, and only the forward component of the effort gets the opportunity to drive the boat. This subject will be touched on again in a later chapter.

Full and Flat Sails: Sails are not flat but, due to their cut, fall into a curve away from the wind. A horizontal, theoretical, straight line from luff to leech forms the chord of that curve while the point of the sail furthest away from the chord is its depth. The fullness or flatness of a sail is gauged from this depth while the position along the chord of the greatest depth determines the sail shape.

A full sail develops more power than a flat one provided it can be kept filled by the wind. But it does not operate as close to the wind as the flat sail and collapses, or becomes 'backwinded', more easily. It does not work so well in stronger winds either, when close-hauled. So what we need is a full sail off the wind and in light winds and a flat sail when beating and in stronger winds.

Dividing the depth into the chord of a sail provides a ratio in which terms the fullness or flatness of a sail can be stated. For a mainsail, a very full cut would be about 1:6 while a flat one would be 1:13. Flatter than this the drive would not be efficient.

Shape and the Power it Develops: A well cut and set

73

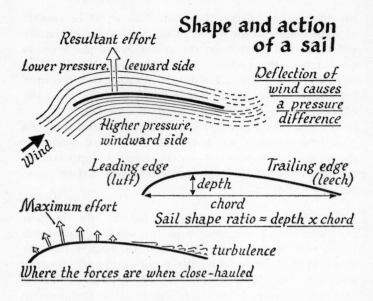

Shape and action of a sail

Resultant effort

Lower pressure, leeward side

Deflection of wind causes a pressure difference

Higher pressure, windward side

Wind

Leading edge (luff)

Trailing edge (leech)

depth

chord

Sail shape ratio ≈ depth × chord

Maximum effort

turbulence

Where the forces are when close-hauled

Aspect ratio

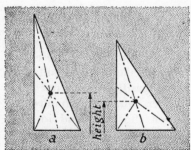

The geometric Centres of Effort are shown on two sails of equal area. High aspect ratio sail "a" will have a better windward ability, but more heeling effect, than low A/R sail "b"

height

a   b

sail has an aerodynamic shape with a maximum depth about a third back from the luff. The amount of power derived from that sail when in use is not constant at all points along that curve, but increases from the luff to a maximum at the greatest depth and diminishes towards

74

the leech. This is particularly so on close-hauled courses but also applies on a reach.

The after parts of the sail curvature, not providing drive on the beat, give increasing amounts of drag. To minimise this and to promote a smooth run off for the 'used wind' leech areas should be as flat as possible.

Cut and Control: Fullness and flow shape is put into a sail in two ways. First by the sailmaker in the way he cuts the panels of cloth and puts in tucks—this is 'tailored flow'. Secondly by setting and such mechanical means as mast bend, kicking strap tightness and other devices—this is 'induced flow'.

Top racing helmsmen may deem it necessary to have full, medium and flat cut sails for light, medium and strong winds, but for club racing or cruising, a medium cut suit, with the knowledge of how to get the best from it, should be enough.

Aspect Ratio: Since most drive is obtained from the area between the leading edge and the maximum depth of the sail, it follows that a taller rig, one with a higher aspect ratio of width to height, is more efficient when sailing to windward. Disadvantages to high aspect ratio for a dinghy rig are a poorer offwind performance and the reduced stability caused by a higher mast and a higher 'centre of effort'.

With boats like the Merlin Rocket, which are given a choice of mast height, the higher aspect-ratio may be preferred for river sailing where efficiency in a fleeting wind supply could be important, and a lower aspect-ratio preferred for sea sailing where the lower centre of effort in the stronger winds could make for easier handling.

Sloop Rig: The intervention of a foresail between the leading edge of the mainsail and the wind causes some interesting things to happen. Apart from creating some drive of its own, the foresail deflects the airstream so that the mainsail requires narrower sheeting angles to the boat's centreline than does the foresail. This is allowed for in sailing boat design. The foresail also creates a slot between itself and the mainsail through which the air must hurry. This effects a pressure reduction which increases the mainsail's pressure differential and gives it more power. A further benefit is that the air flow is smoothed on the leeward side of the main and the creation of harmful eddies is reduced.

The Size of the Slot: The size of the slot between the foresail and the main is critical to windward performance. Too narrow and the air cannot get through and the mainsail is backwinded; too wide and the lee of the mainsail does not get the benefits of successful deflection. A number of factors are involved with this.

The most obvious is the athwartships position of the fairleads. With sliding fairleads which are adjustable from side to side any slot faults are easily corrected but the movement of fixed fairleads should be well considered, for wind strength is also a factor affecting the optimum setting. A slot which is right for light winds may become too narrow in stronger winds when more air tries to get through.

The fore-and-aft position of the fairlead can affect the slot. Too far forward and the foresail has too much flow, deflecting the air and backwinding the main; too far aft and the foot is tight and the leech slack, allowing it to sag off to leeward and opening the slot. Cures for this are noted under 'setting the foresail' (p. 85).

Alterations to mast rake may affect the slot width.

Any movement aft tends to close the slot while the converse forward movement opens it. Such movements also affect the fore-and-aft sheeting position for the foresail.

Shroud tension can also play a part as can the bending

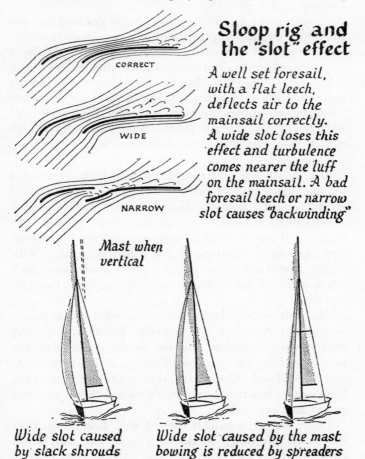

Sloop rig and the "slot" effect

A well set foresail, with a flat leech, deflects air to the mainsail correctly. A wide slot loses this effect and turbulence comes nearer the luff on the mainsail. A bad foresail leech or narrow slot causes "backwinding"

CORRECT

WIDE

NARROW

Mast when vertical

Wide slot caused by slack shrouds

Wide slot caused by the mast bowing is reduced by spreaders

77

characteristics of the mast itself. Slack shrouds allow the mast and hounds to sag off to leeward, giving a wider slot, while a flexible mast 'bowing' to windward opens the slot in two ways—by virtue of the bow and by shortening the distance between step and hounds and causing slacker shrouds.

Foresail size has a direct effect on the slot and a genoa jib, which overlaps the mainsail to a considerable extent and causes excessive backwinding, may give a dinghy a poorer windward performance than a smaller, but more efficient 'working jib'. But for the sake of the genoa's better offwind performance this may be acceptable. For strong wind sailing, however, a smaller jib, or one cut with a slightly concave leech which will maintain an efficient slot, can be a useful additional sail.

Sailcloth: Canvas and cotton has now been superseded by man-made fibre cloths. These have advantages of greater density, imperviousness to shrinkage or stretch when subjected to wet and dry conditions, resistance to mould and (in the case of polyester, which is commercially known as Terylene in this country) 'rigidity' with which to hold sail shape. Nylon is a man-made yarn which has stretch and its cloth suits the requirements for spinnakers.

Slight drawbacks include degradation from exposure to sunlight (which is not a major problem to dinghy sails which are seldom exposed to the extent that the sails of cruising yachts are) and the vulnerability to chafe and wear of the sail stitching, due to the fact that in the hard cloth the stitches are unable to bed down, so stand proud and exposed.

The material is quickly melted with heat. Sometimes this is an advantage in that it can be cut with a hot knife which at the same time fuses the severed ends to prevent their fraying, but it is a misfortune when a carelessly held cigarette results in a neat round hole.

The rigidity of Terylene cloth applies along the thread-lines only. The cloth accepts stretching from a pull only a few degrees away from these lines, which results in a

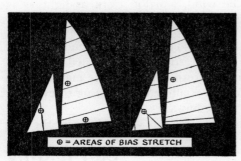

⊕ = AREAS OF BIAS STRETCH

Sail cuts

Cloth "stretches" and imparts flow if the panels meet the sail edges at an angle

"Tailoring" the abutting panels and curving the "straight" edges "builds in" an amount of fullness

gathered fold. When designing and cutting a sail, a sailmaker arranges the panels of cloth so that tensions will be accepted without stretch along the leech, so keeping that area flat, but where a controlled amount of stretch helps to induce flow, along the luff and foot, the panels meet the edges at a slight angle.

A plastics coating, or plating is put on to the woven material during manufacture which gives it smoothness and added density. This cannot be replaced after the sail has been made up. Frequent careless refolding can damage this surface.

The material can be obtained in various weights but for a dinghy's working sails the $5\frac{1}{2}$ ounce is standard.

This weight is per the square yard but probably metrication will soon be upon us. A heavier weight of cloth is sometimes used for heavy-weather sails or parts of a sail which need to be more self-supporting, the roach area and the head, while for spinnakers, sails which are supported only by the wind, a nylon cloth as light as $1\frac{1}{2}$ ounce can be used. (See Sailcloth weight, p. 100.)

Care of Sails: While new Terylene sails do not need the careful breaking in and stretching accorded to cotton sails, pulling them progressively out to the marks during the first few settings does allow the stitching to even out and settle along the seams.

If you are leaving sails for some time in the bag

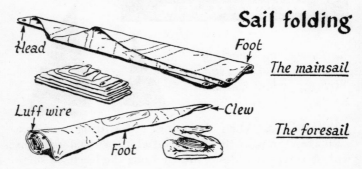

Sail folding

The mainsail

The foresail

between outings they should be washed free of salt and grit and dried before folding. Being left in the boat stuffed under the foredeck or rolled around the boom is all right for short intervals, but it will shorten the life of a sail if done frequently.

Folding the foresail is a matter of rolling into a tube of about a hand's span diameter along the luff wire, or folding zig-zag fashion if the luff is wireless or roped, and then rolling. Plastics windows, although possessing great resilience, should not be creased.

Mainsail folding can be done either along the length or across the width. Lengthwise, leech is folded to luff and then halved. The long wedge shape is then folded and heaped in zig-zag folds to form a neat parcel. Alternatively, the luff is divided by folding several times and a zig-zag folded or rolled parcel made of the resulting shape.

At the end of the season the sails should be inspected and possibly sent back to the sailmaker for washing and repair. If you like to do your own maintenance, the drill is to wash, dry, inspect, repair and store lightly folded in a dry airy place. Grease spots can be removed with cotton wool and lighter fuel, jelly-solvent, or carbon-tetrachloride but it would be as well to try such fluids on the sailbag first, if that is of similar material, to note the effect. For the wash a little detergent may be added to the water and the sails well rinsed afterwards.

Broken and chafed stitching should be stitched over, using polyester yarn and following the original stitch holes. Small tears can be patched if you can get some identical cloth. To make a patch cut out a rectangle of cloth larger than the tear, turn under a fold on each side and stitch to the sail. Next cut four diagonals from the centre of the tear so that the jagged edges can be folded under between the patch and the sail. Finally, stitch these four edges to the patch.

A temporary repair can be made with the use of self-adhesive tape.

The Foresail: Panels of cloth are usually arranged so that they run parallel either to the luff or the leech or to both in the case of a sail with a mitre seam. This allows the strains from the sheet to be taken along the thread lines and the leech area to be kept flat—most important for a good run off to the main.

The sailmaker cuts the luff to a shallow 'S' which, when pulled tight by the halyard, puts the flow into the sail. The foot may be straight or rounded and given a few vertical tucks to put in some extra flow.

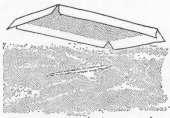

Sail patching

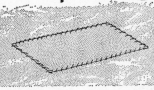

The tear and the patch

Patch stitched with raw edges turned under

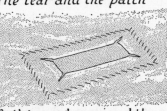

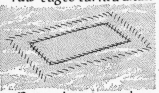

Sail turned over and the tear cut back neatly

Raw edges turned under and stitched

The luff is usually wired or roped to take the halyard strain and to ensure the very tight leading edge which is necessary for good windward performance. This may be stitched to the sail or may be enclosed in a pocket and attached to it only at the head. An adjustable strap or a control line at the foot can then apply varying amounts of tension to give differing amounts of flow.

Jib hanks, plastics or piston type, attach the luff to the forestay not so much for support but more for guidance in raising and lowering. The tack is attached to the stem either directly with a shackle or elevated by a wire strop. A refinement can be a swivelling shackle which eliminates any twists in the luff.

The Bermudian Mainsail: Panels of cloth are usually laid approximately at right angles to the leech so that strains are taken from clew to head. They meet the luff at an angle which permits considerable stretching by the opposing tensions of the halyard and the gooseneck downhaul. The leech is frequently given an outward curve and this area, between the edge and a straight line between head and clew, is known as the roach. This is supported by three or more battens and there may be a leech line, a draw cord threaded through the tabling, or hem, at the edge.

The sail is given flow by the curved cut at luff and foot which, when set upon straight spars, produces the shape. There may also be shaping in the panels and horizontal tucks in the lower part of the luff. Foot and luff have bolt ropes which may slide into mast and boom grooves or be given sliders to engage with tracking. A simplified rig may have a pocketed luff which slides over mast or boom or be loose-footed, that is, attached to the boom only at tack and clew.

A headboard of aluminium or plastics spreads the pull of the halyard and there may also be some stiffening at the clew, the adjustment of which can be a simple tie or a running tackle which returns inboard so that alterations can be made under way.

A common refinement is the Cunningham hole device which is a means of pulling a bunt of sail down from the luff area to improve the flow of the sail for close-hauled sailing. The bolt rope is elasticated and a control line from the boom, runs up through a grommet in the luff area of the sail just above it and then down to a cam cleat on the mast. Alternatively, there is a Jack Holt

Flow control

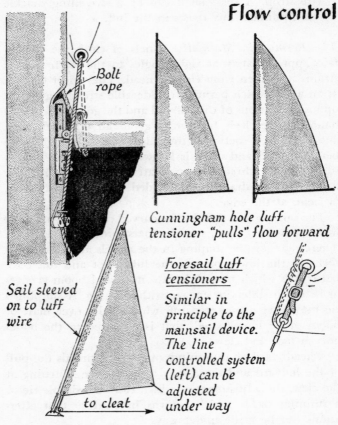

Bolt rope

Cunningham hole luff
tensioner "pulls" flow forward

Foresail luff
tensioners

Similar in
principle to the
mainsail device.
The line
controlled system
(left) can be
adjusted
under way

Sail sleeved
on to luff
wire

to cleat

device which clamps the bolt rope itself for hauling on.

The Gunter Mainsail: Although similar in shape and
appearance to the Bermudian mainsail to the casual
observer the leading edge is carried on two spars, the
mast and the yard, or gaff. The mast edge is still called
the luff but above the throat, which is at the position

of the gaff jaws, the edge is known as the head, while the top of the triangle becomes the peak.

Construction of the sail follows that of the Bermudian with panels running at right angles to the leech but the leading edge is usually cut straight as the slightly out-of-line attachment of mast and yard gives the necessary flow which can be augmented by control of the tension of the foot.

Bending (attachment) to the spars may be by lacing, by lacing and groove, or by track and sliders. The gaff jaws act as the sliding hinge for the yard the cheeks of which are 'closed' by a parrel of wooden beads or a simple tie of lacing or shock cord.

The foot of the gunter sail is more likely than the Bermudian to be loose footed. This is on account of the fact that the sail is usually flatter and this arrangement allows more flow to be put in.

Setting the Foresail: Set up the luff and bowse the halyard down on its cleat to give a tight luff. A pull, by a friend on the forestay, with some jib hanks temporarily disconnected, is an aid to doing this. Care should be taken, if the halyard is internal to the mast, to see that the end does not disappear into the sheave-box and up the mast. The usual way is to tie a stopper knot, such as the 'figure of eight' in it, another is to tie the ends of the foresail and mainsail halyards together with a reef knot.

Mechanical aids to a tight luff include a winch, a tackle and a tensioning lever. The latter embodies the Highfield principle of a hook on a lever on to which a loop from the wire halyard is engaged, then to be tightened when the lever is pushed down and which remains in the down position when its hinge point is in line with or past the direction of pull.

This tightness not only prevents any sagging off to

leeward but it draws the flow in the sail forward to the area close to the luff and enables the sheeting to be effective in flattening the leech.

Sheeting position of the fairlead can now be tested. For a good windward setting both leech and foot need to be under moderate tension. Adjustable fore-and-aft slid-

Sail controls

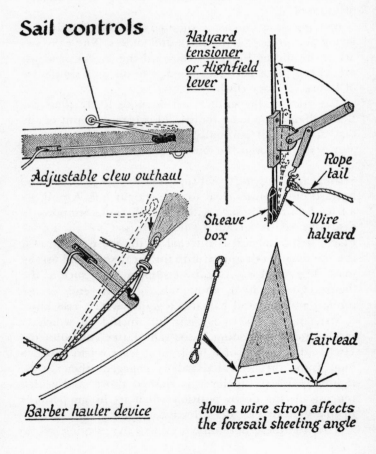

Adjustable clew outhaul

Halyard tensioner or Highfield lever

Rope tail

Sheave box

Wire halyard

Barber hauler device

Fairlead

How a wire strop affects the foresail sheeting angle

ing fairleads simplify this positioning. Too far forward gives excessive tension to the leech which, at its worst, results in a folded or cupped leech, and too much flow, both disastrous to the slot effect. Too far back and the foot may develop a fold and the leech become floppy. With a rounded foot to the foresail there may still be a fold when the sail is correctly sheeted and drawing and it may even 'motor'; that's to say it may flutter noisily. But this is not necessarily harmful and the rounded foot will pay benefits on the offwind courses.

On boats without fore-and-aft adjustable fairleads the foot and leech relative tensions can be adjusted by raising or lowering the tack by means of shorter or longer wire strops. Be prepared for slight readjustments when the mainsail is also set and when actual working conditions afloat manifest themselves.

Setting the Bermudian Mainsail: With the foot of the sail set on the boom, the tack secured, the sliding gooseneck loose and the battens inserted in their pockets, the luff is guided into the mast luff-groove (or the sliders are engaged) and the sail is hoisted. When the head is level with the upper black band on the mast the halyard is cleated and extra tension given to the luff by down-hauling the boom with the sliding gooseneck and screw lock, but not lower than the bottom black band on the mast. Amounts of tension will vary with wind conditions; less when it is lighter and vice versa. A halyard winch or a tensioning lever, similar to that for the foresail, can adjust this tension by the halyard.

The clew outhaul can then be adjusted and again tension depends on wind conditions. Among other factors influencing the set of the sail are the kicking strap tension (see page 121) and the pull of the mainsheet, particularly if this is a centre-attached one and the boom is at all bendy.

A sail can be set on shore or at the mooring but the only true judgement of it is to view it on the water and in use. There other tensions are being applied, not least the pressure of the wind which tends to 'blow' any forward fullness further aft.

The sail that sets without creases is a rare beast. Radiating creases from the corners are fairly harmless but a pronounced crease from clew to head should be eradicated.

Easing the clew outhaul may do this, unless there is a fault with the battens. Tight ones transmit tensions into the sail and a slight shortening may be needed. Lack of flexibility is another batten fault and the inner ends, if they are of wood, can be shaved down so that they mould themselves into the curve of the sail.

Controlling Sail Shape Afloat: The sail is initially set for windward sailing, as this is invariably the first leg of a racing course. Halyard tension induces some flow shape into the forward part of the sail, but, apart from wind pressure, other influences can work to reduce this too much. Mast bend is one of these, bending forward at its mid-height under the combined pressures of the mainsheet (a backward pull transmitted to the masthead) and the kicking strap (a forward push at the gooseneck). One solution is to use the Cunningham hole gear, another is to limit mast bend.

For keel-stepped masts which pass through a mast gate, some restriction can be made at deck level, by the insertion of dropped in chocks, but this will not restrict mast bend higher up the mast. For this limited-swing spreaders are required. Basically aluminium rods between mast and shrouds, their swing is limited in their brackets and this stop comes into operation as the mast bends forward.

Spreaders

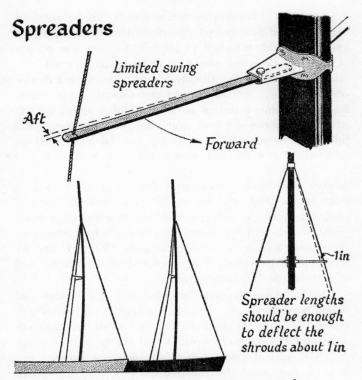

Limited swing spreaders

Aft

Forward

~1in

Spreader lengths should be enough to deflect the shrouds about 1in

Free swinging spreaders (left) limit lateral bend but not forward bowing.

Limited swing spreaders (right) do restrict this which retains fullness in the luff area of the sail

On reaching and running courses the kicking strap can be eased slightly. Not too much, as its function is to stop the boom lifting, but enough to prevent it causing any side bend of the mast through the gooseneck.

A slack luff to the foresail is advantageous to off-wind sailing but this is not easily arranged unless the halyard is hooked to a tensioning lever, which can be released and tightened as required. Another sophisticated method has the shrouds attached to Highfield levers, the release of which enables the whole rig to be tilted forward.

The 'Barber hauler' (illustrated on page 86) is a device which aids the soft setting of the foresail on a reach. Basically a line attached to a ring which slides on the jibsheet between fairlead and clew, its purpose, when tightened, is effectively to bring the fairlead position forward, to give the sail more curve.

Setting the Gunter Mainsail: The head of the sail is bent on its yard, secured at the peak and the battens inserted. A single halyard may be attached to a fixed point on the yard but if the attachment is to a wire span there will also be a throat halyard. The tack can be attached to the boom, if loose footed or, with a tracked or grooved boom, fully fitted.

The gaff jaws can now be presented to the mast and closed by parrel or tie then, as the yard is partly hoisted, the luff is laced to the mast or the luff sliders engaged. The yard is then fully hoisted and the halyard cleated. A fairly vertical yard helps windward performance and imparts flow into the forward part of the sail. Extra tension may be given by a sliding gooseneck or a tack downhaul working through a grommet in the luff. The clew outhaul can now be adjusted to suit wind strength, to eliminate creases or, in the case of a loose footed sail, to give the sail some additional flow.

Easing the halyard to give the yard a slight rake (simulating the bend of the Bermudian mast) takes out some of the forward flow and distributes it further aft. The consequent slackening of the leech can then be taken up by tightening the kicking strap.

RIGGING ADJUSTMENT AND TUNING

Most dinghy rigging (two shrouds and a forestay) has bottle screws at the anchorage for length and tension adjustment. A simpler arrangement is the lanyard of pre-stretched Terylene cordage.

To check mast rake the boat is set up level ashore, with the aid of a spirit level set along and across the top of the

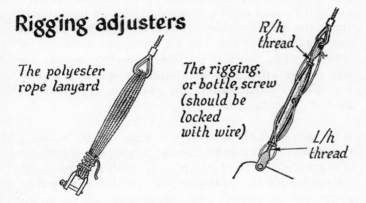

Rigging adjusters

The polyester rope lanyard

R/h thread

The rigging, or bottle, screw (should be locked with wire)

L/h thread

centreboard case, or thwart, and by chocking stem or stern and one bilge or the other. A weight is hung from the main halyard down into the boat and, without wind interference, a vertical mast is achieved by adjusting the shrouds against each other. The forestay is then lengthened or shortened to give a reasonably taut set of wires—but not so taut that excessive compression stresses are set up in the mast, with consequent risk of bowing or distortion.

The mast should be raked aft with the weighted halyard hanging free in the boat (or forestay should be eased and the shrouds tightened until this is so). The amount of rake can be measured at deck level from the halyard to a fixed part of the boat forward of it—probably the deck. This is your datum distance and should be remembered or noted.

Performance under sail is the next thing and there is nothing like a 'trial horse' to check this, although racing conditions will suffice. If windward performance is poor the boat can be set up again and, by adjusting the forestay in relation to both shrouds, the mast raked further aft. If the downwind performance is the poor one the mast should be raked further forward. Never be half-hearted about the initial tuning adjustment, but alter the rake considerably. Only this way will what you are doing be obvious. True, you may get some bad performances one way or the other, but by starting thus and gradually narrowing the parameters the optimum adjustment is likely to be achieved.

Wire Rope: This is used for most dinghy standing rigging and sometimes for running rigging and low stretch is its great virtue. Solid wire is occasionally used for standing rigging for heavy dinghies which seldom unstep their masts and for jumper and diamond staying on racing dinghy masts. Its virtues are stability, smoothness and low windage. Its thickness is measured either by the diameter in millimetres or by Standard Wire Gauge.

Rope for standing rigging is used popularly in two constructions—7 × 7 and 1 × 19. The former is made up of six 'ropes' each of seven twisted strands laid around a central rope. This is less smooth than the other construction which has a central core of one wire, surrounded by six more and then an outer layer of twelve. Running rigging is mostly of 6 × 19 or 6 × 7 construction.

Wire is made of either galvanised or stainless steel. Galvanised has a shorter life than stainless but, for standing rigging, can be dressed with boiled linseed oil during lay-up, which dries and protects it. Stainless steel wire has an oxide film on rustable steel wire which can be broken down if subjected to long periods of dirt and wet so the lesson is, keep clean for long life. A third method for the protection of wire is a plastics covering which is all right until the sheath is damaged, water penetrates, and deterioration spreads underneath.

Splicing of thimbles into the ends of rigging is by swaged ferrules put on with a special cramping tool. For electrolytic reasons aluminium ferrules are used for galvanised wire and copper ones for stainless steel.

'All wire' halyards cannot be tightened by hand but require either a tensioning lever or a winch. A rope tail to a wire halyard makes it amenable to hand cleating and the illustrated long splice is the one to join the two halves.

WIRE ROPE BREAKING STRESSES

Standing Rigging:

Dia		1×19	*Stainless*	7×7	*Stainless*	7×7	*Galvanised*
mm	in	kg	lb	kg	lb	kg	lb
2	5/64	300	650	250	550	280	610
2·5	3/32	430	950	360	800	–	–
3	1/8	770	1700	610	1350	630	1380
4	5/32	1250	2750	920	2050	1100	2420

High resistance to corrosion: minimum stretch: hand splicing difficult.	Fair resistance to corrosion: more stretch and flexibility than 1×19 construction; hand splicing easy.

For most medium-sized dinghy rigs the 3 mm size is suitable. This can also be used for kicking-strap purchases and wire strops for foresails.

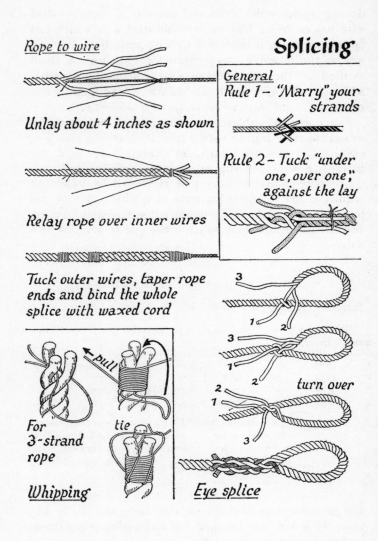

Rope to wire

Splicing

Unlay about 4 inches as shown

<u>General</u>
Rule 1 – "Marry" your
    strands

Rule 2 – Tuck "under
    one, over one,"
    against the lay

Relay rope over inner wires

_Tuck outer wires, taper rope
ends and bind the whole
splice with waxed cord_

pull

For
3-strand
rope

tie

turn over

<u>Whipping</u>

<u>Eye splice</u>

Running Rigging:

Dia		Stainless			Galvanised		
		6×7	6×19		6×7	6×19	
mm	in		kg	lb		kg	lb
2	5/64	Slightly	–	–	Slightly	–	–
2·5	3/32	less strong	300	650	less strong	–	–
3	1/8	than 7×7	590	1300	than 7×7	500	1100
4	5/32	(see above)	900	2000	(see above)	880	1940
		which is			which is		
		also used			also used		

These ropes are formed around a PVC, or hemp, core, which gives greater flexibility but more stretch and less resistance to corrosion than the 'all wire' counterparts have. They can also be crushed over a small diameter sheave. Hand splicing is easy for the 6 × 7 constructions. For most dinghies the 2·5 mm size is suitable for halyards. A larger size may be preferred for winches (see p. 115).

ROD WIRE BREAKING STRESSES

Stainless, used for jumper and diamond stays on the mast.

| Dia | Strength | |
mm	kg	lb
1·5	270	600
2	450	990

NATURAL AND SYNTHETIC ROPE

This is measured by its circumference in inches or by its diameter in millimetres. Natural fibre ropes have not been in demand by dinghy sailors since the introduction of synthetic fibres with their advantages of similar wet and dry strengths, minimum water absorption, length stability in dry and damp conditions and resistance to rot and fungal attack. The one disadvantage of greater cost is offset by longer life.

Cotton Rope: Much appreciated as sheets for its handling qualities but with great water absorption. Now superseded by plaited polyester.

Sisal: A cheap general purpose rope with little water absorption but rough to handle.

Hemp: Makes cheap halyards, lacing or reefing line. May be purchased lightly tarred which waterproofs it.

Manilla: The best quality natural fibre, the strongest and the easiest on the hands. Suitable for halyards.

Polyester rope: Marketed as Terylene in this country. Has great strength and low stretch properties, especially the pre-stretched 3-strand which is good for halyards. Plaited rope is ideal for sheets. Rather expensive.

Nylon Rope: Cheaper than polyester with even more strength but loses a slight amount when wet. Its elasticity makes it suitable for certain uses, such as an anchor cable.

Polypropylene Rope: A cheaper synthetic rope than the other two and slightly less strong. Its lightness makes it buoyant in water. It is supple and has applications as mooring ropes, tow lines and painters.

Colour: Synthetic rope can be obtained coloured for the purposes of recognition on the cleat. The suggested colour code proposed by Marlow Ropes is as follows:

Red—Spinnaker; Blue—Genoa; Gold—Jib; White—Main/General.

WHIPPING

All natural fibre ropes must have whipped ends to prevent fraying. There are many types of whipping from the sailmaker's, which is made with a needle, to a simple seizing with whipping twine. I illustrate (on page 94) one which is easy to do and gives reasonable service.

Three-strand synthetic rope can also be whipped similarly, fitted with a Marlow patent whipping tube, which shrinks under heat, or simply 'heat fused' with a match. The molten end can be shaped with a damp cloth, which carries less risk than the damp finger and thumb technique. This also goes for plaited rope.

ROPE SPLICING

This is a useful, but not essential craft for dinghy sailors. Most applications will be found for the eye splice (illustrated on page 94), which can be formed with or without a thimble inside.

KNOTS, BENDS AND HITCHES

A dinghy sailor can get by with very few but he must learn at least five which I note in order of the frequency that you will probably use them. *The figure of eight*—the stopper knot which is made at the ends of sheets and halyards to prevent them disappearing into sheave boxes and through fairleads. *The round turn and two half hitches*—used to secure the painter to a ring or a post. *The bowline*—a non-slip knot, yet easily untied for attaching the painter to the dinghy's mooring cleat or

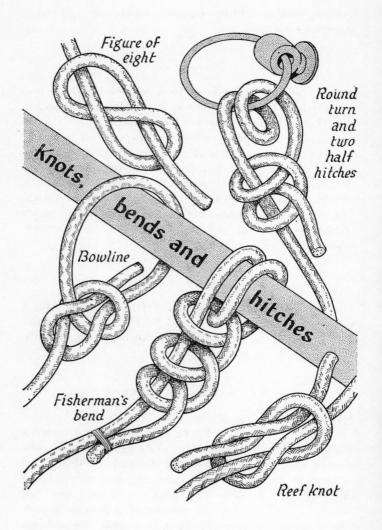

Figure of eight

Round turn and two half hitches

Knots, bends and hitches

Bowline

Fisherman's bend

Reef knot

similar positive attachments. Two unequal sized ropes can also be linked each with a bowline. *The reef*—for tying reefing points, two equal sized ropes together and many other applications. *The fisherman's bend*—a more permanent attachment (and more difficult to undo) which is used for tying the warp to the anchor. After the last half hitch the loose end should be seized to the standing part.

SUGGESTED CHOICE OF ROPES AND SIZES

Ropes are chosen more for their handling, flexibility and non-stretch qualities than for their strength, although this must be considered, as the strengths of the ropes which 'look' and 'feel' right are usually more than adequate.

Sheets: For most dinghies $1\frac{1}{4}$ in circ (10 mm dia) soft plaited 8-strand polyester or polypropylene. An alternative is the same size and material in 3-strand hawser laid.

Halyards: 'All rope' halyards can be formed with $\frac{3}{4}$ in circ (6 mm dia) 3-strand hawser-laid 'pre-stretched' polyester. The same material can be spliced to wire to combine the minimum stretch of wire with the handling quality of rope.

Anchor Cable: (See 'Anchoring and Mooring', Chapter 8).

Spinnaker Guys: Polyester, polypropylene or nylon, preferably plaited, for flexibility and lack of 'twist', in $\frac{1}{2}$ in circ (4 mm dia) or $\frac{3}{4}$ in circ (6 mm dia).

99

Painter: Polypropylene or polyester 3-strand hawser laid 1 in circ (8 mm dia).

Kicking Strap: ¾ in circ (6 mm dia) polyester, either plaited or hawser laid.

Lengths: These must be adequate and take into account, for example, the number of parts in a mainsheet purchase. Each additional block increases the length required by the distance between the inboard block and the one on the boom at the limit of its swing.

SAILCLOTH WEIGHT (*see page 80*)

Metrication is indeed upon us and there is also an American Standard yard (which means a yard of cloth 28½ inches wide as distinct from the square yard used for the British Standard), so care must be taken when ordering sails. Thus a 5 ounce cloth to the British Standard is a 4 ounce cloth to the American Standard while under metrication this equates with 170 grams per square metre.

Sailcloth Conversion Table

British Standard (ounces per square yard)	American Standard (ounces per yard 28½ inches wide)	Continental Standard (grams per square metre)
1	·7	34
1·5	1·2	51
2	1·5	68
2·7	2·2	94
3	2·5	102
4	3·2	136
4·5	3·5	153
5	4	170
6	4·7	204
7	5·5	237

5

Masts, Spars and Ancillary Equipment

MASTS AND SPARS

Wooden Mast: The traditional mast material (spruce, ash, or fir) still has its adherents who, in a world with an increasing shortage of timber of the right kind and quality, will find their allegiance more and more difficult.

Advantages include buoyancy, easy shaping and tapering for obtaining certain bend characteristics, a fair resistance to impact, the acceptance of screws for fittings and relative cheapness.

Drawbacks may be those of maintenance, uncertain strength characteristics for a given size and section and some length instability in damp and dry conditions which affects tuning.

Wooden Spars: With the simpler spars such as a yard, bowsprit or boom, wood is a more doughty contender. Apart from the easy fixing of screws and fittings, it allows the stiffness of wood to be put to good effect. An alloy mast is often coupled with a wooden boom which can be given a deep and narrow section for minimum bend, so that kicking strap or the centre mainsheet tension is transmitted to the mast to promote bend there.

Aluminium Alloy Mast: Advantages include a 'designed

bend' characteristic, minimum maintenance, especially when anodised, and lightness.

Drawbacks can be listed as, a liability to denting and to bending across the narrow diameter of its section, resulting in breakage; negative buoyancy, unless foam filled or with sealed buoyancy compartments; and the need for a riveting tool for the addition of fittings.

It is worth noting that non-anodised alloy suffers no deterioration through corrosion, which can be removed and the surface restored by the occasional application of polish. Also, that broken spars can be welded and repaired.

Aluminium Alloy Spars: This material has a lightness advantage over wood as a main boom as well as maintenance benefits. Being hollow it is possible to lead the adjustable clew outhaul lead forward to a handling position or a winch near the mast. Flexibility is to some extent overcome by an oval section and internal webbing and, in the case of a centre mainsheet, the take-off point is spread between two or more take-off points for blocks separated along the boom. Flexibility is not a good thing in a boom as it flattens the foot area of the sail, in the case of a centre mainsheet, and allows an upward curve in the centre, in the case of the transom mainsheet.

Stainless Steel Mast: The main advantage is a greater 'strength to section' ratio than possessed by either wood or alloy, which enables thinner sections to be used. Since this means less windage this is of importance in racing. Other selling features are those of minimum maintenance and denting resistance.

Not such good features are the lack of buoyancy and the difficulty for the amateur, of fixing fittings, while the cost is greater than either alloy or wood.

Other Materials: In the past bamboo has been used on some local one-designs for both mast and spars. It has strength and lightness but lends itself more to laced-on sails than to modern tracks and fittings. What the future holds in store is carbon fibre: this has great prospects but is likely to be more expensive than any other material.

Amateur Spar Construction: All three of the popular spar materials are suitable for the do-it-yourself yachtsman. For wood he needs some skill, particularly for a hollow mast, and woodworking tools, including access to many clamps. For alloy he is limited to making spars from round extrusions with track and fittings and will need access to a riveter. In stainless steel he can obtain kits for completion, which will save him a little money. Kits are also available for aluminium masts.

Square-section Mast: A simple solid mast for solid wood construction can be laminated from half inch board with opposing grain to reduce warping. Its flat side makes a good seating for track which, with suitable fittings, will take both the sail sliders and a gunter yard.

Round-section Mast: The usual choice is solid wood for lugsail dinghies. This is also used for the gunter rig as it has the toughness to stand up to the battering and sliding of the gaff jaws.

Many alloy masts have a round section which is much cheaper than the specialised sections. Stainless steel is also made in round section and is given more fore-and-aft bend resistance by the mast track which is riveted on from gooseneck to truck.

Some smaller dinghies have an unstayed, round section, wood or alloy mast, stepped on the hog and passing through a hole in the thwart or the deck for support.

Mast sections

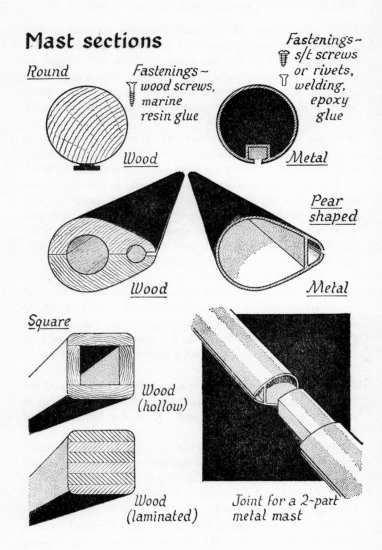

Round

Fastenings ~ wood screws, marine resin glue

Wood

Fastenings ~ s/t screws or rivets, welding, epoxy glue

Metal

Pear shaped

Wood

Metal

Square

Wood (hollow)

Wood (laminated)

Joint for a 2-part metal mast

Pear-section Mast: This apparently streamlined shape is the result of the integration of the mast's round section with the luff groove and has the advantage of giving more fore-and-aft stiffness with which to withstand close-hauled stresses. When the boom is squared off on a reach the pull is across the narrower dimension and bending may have to be reduced by spreaders or diamond staying.

This section is used by both wood and alloy masts. In wood, weight is saved, and with internal halyards windage is reduced, by making the mast hollow. Relative to a solid mast some strength is lost but reduction in top hamper gives the boat a gain in stability. The mast is made in two longitudinal halves to enable the hollows for the halyards and the luff groove to be formed and then made integral by the strength of a synthetic resin glue joint. Tapering towards the top effects a further saving of weight and influences the bending characteristics.

In alloy, weight saving is even more marked. Diameters can be made smaller (with the same strength as a wooden mast so saving on windage) and the standardised nature of the material means that both strength and bend characteristics are factors that can be more accurately specified.

Due to manufacturing limitations extrusions must be parallel along their length but, by cutting out a long triangle of metal at the top and welding the edges, a sort of taper is formed. Welding is used again in the fitting of a lateral internal web, which closes off the luff groove (and space for the internal halyards) from the main internal diameter which may be foam filled or made air tight to provide buoyancy. The web, or webs, also give stiffness or the required bending characteristics.

Rotating Masts: The amount of windage on a mast is

not insignificant, as can be felt during the stepping of one in a fair wind, but the smaller the diameter the less it will be. So stainless steel masts, with their minimum diameter, are best on this count. Round masts in wood or alloy are next best and the apparently streamlined pear-shaped section masts come equal last with square section masts. The explanation for this is that, when close-hauled, the section is presenting much of its major diameter to the wind.

The rotating mast minimises this. The mast step is cupped, the attachment point of the shrouds to the hounds is on the forward face of the mast and the mast rotates and lines up with the sail by the pressure of the boom. There is a supplementary benefit in that the leeward side of the mast is more in line with the leeward side of the sail and so causes fewer eddies.

The unstayed mast, as in the case of the Finn and OK dinghies, can be made to rotate by the boom being tenoned into the mast.

Two-part Masts: One of the snags with the long Bermudian mast is the problem of carriage for trailing. In alloy this is answered by the two-part mast which stows inside the boat and telescopes together with a collar and spigot. Since this puts extra weight aloft and an uneven bend characteristic in the mast, this is not for the highly competitive racing classes.

Mast Stepping: The task is simplest with an unstayed mast, which is set up and locates in a hole in the thwart or foredeck and rests in a step on the hog. In the smaller sizes of Bermudian stayed masts, or for gunter or lugsail rig short masts, it is not much more difficult. The mast is set up with shrouds and forestay attached at the hounds and all that remains is for these to be adjusted and fixed

Rotating masts and windage

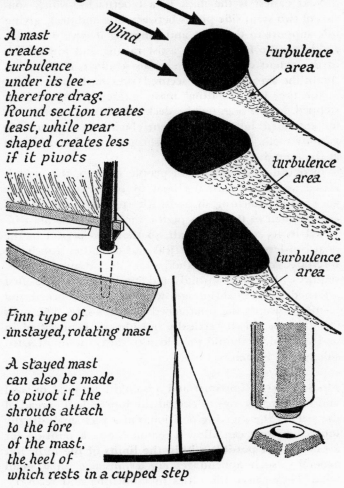

A mast creates turbulence under its lee – therefore drag. Round section creates least, while pear shaped creates less if it pivots

Wind

turbulence area

turbulence area

turbulence area

Finn type of unstayed, rotating mast

A stayed mast can also be made to pivot if the shrouds attach to the fore of the mast, the heel of which rests in a cupped step

to their respective chain and bow plates.

Next easiest is the mast in a tabernacle—which consists of two stout side pieces between hog and deck, giving side support to the mast and hingeing it with a bolt. A tackle on the forestay can assist raising and lowering it to the extent of being able to do it afloat, if necessary. Again the shrouds can be secured afterwards quite easily.

For the 'free standing' mast, gunter or Bermudian, stepped on the hog, on the deck or midway between on an extended horn of the centreboard case, there are various methods. In all, the shrouds are first attached at the hounds.

In one method, with two people to help, the mast is raised and lifted into the boat by one while the other goes round attaching shrouds. Alternatively, the shrouds are attached to the chain plates, the heel of the mast inserted in its step leaning aft, so that one person in the boat can lever it to the vertical and the other attach the forestay. A further method, for deck-stepped masts, requires that one shroud and the forestay are first attached to their anchorages with the boat chocked and leaning towards the unattached shroud's side. The mast can then be raised vertically outside the boat, lifted in and stepped. It should then lean towards the unattached side, which is then secured.

Minor Spars: A bowsprit and a bumkin are used in yawl and some lugger rigs to extend the fore-and-aft spread of the sails or the sheeting position. On a very small dinghy too, a bowsprit can allow a reasonable tack position for a foresail not possible within the limits of the hull. But generally, such appendages are a nuisance on a small boat. They cause difficulties in launching and recovery, mooring, (both to a jetty and to an anchor) and when used as a tender to a larger vessel.

Mast stepping

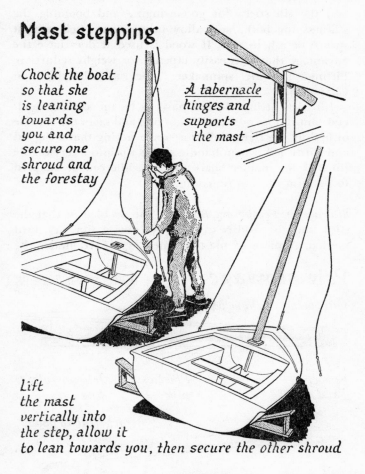

Chock the boat
so that she
is leaning
towards
you and
secure one
shroud and
the forestay

A tabernacle
hinges and
supports
the mast

Lift
the mast
vertically into
the step, allow it
to lean towards you, then secure the other shroud

Rigidity is their prerequisite and they are invariably made of stout wood. For additional support the bowsprit is braced by a bobstay to the forefoot and the bumkin by lateral wires to the transom.

Carried spars include the spinnaker boom (or **pole**)

and the jib stick—for goosewinging and booming the stiffness for both, and alloy tube counts well against spruce or ash in this. If wood is used it does have the advantage that it is easily tapered for weight reduction. Fittings for the spinnaker boom are described in Chapter 9.

Jib stick fittings do not have to be the same at both ends but it does no harm if they are, and saves the trouble of finding the right end. It is worth noting that a shaped and varnished broom handle makes up into a serviceable jib stick for smaller boats and a paddle can be adapted to perform the two functions.

Stowage of Temporary Booms: The problem is that the stick must be readily to hand yet out of the way until needed. A piece of plastics water pipe fixed under the

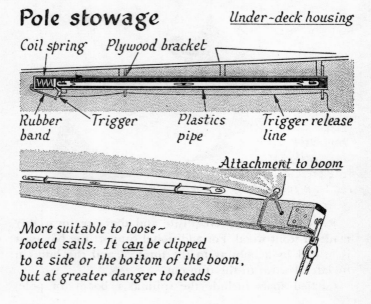

Pole stowage   *Under-deck housing*

Coil spring   Plywood bracket

Rubber band   Trigger   Plastics pipe   Trigger release line

Attachment to boom

More suitable to loose-footed sails. It can be clipped to a side or the bottom of the boom, but at greater danger to heads

foredeck, or sealed into a forward buoyancy tank at its bottom end around the lip is one answer. It can be spring loaded to eject the pole by a trigger mechanism.

Clips under the side decking, on the floor or alongside the centreboard case offer alternatives. On some boats it may be expedient to fit the clips along the top of the foredeck, where the boom is certainly to hand, while on boats with a loose footed mainsail, as with the Mirror dinghy, clips can be set on top of the main boom so that the smaller spar can be sited there.

SHACKLES AND BLOCKS

Shackles: The simplest accessory is also one of the most important as it holds much of the rigging together. Of course, the shackle must have adequate strength for the job it has to do and, as strength increases with size, the shackle must be exchanged with a bigger one if it fails. But it is better to have it right to begin with. This means that the breaking strain of the shackle should be at least equal to that of the wire or rope it holds.

'D' Shackles, Breaking Loads

Pin dia		Die cast manganese bronze		Drop forged galvanised		Stainless steel strip	
mm	in	kg	lb	kg	lb	kg	lb
3	1/8	406	896	280	616	364	800
5	3/16	763	1680	840	1848	908	2000
6	1/4	1120	2464	1636	3584	1726	3800

This is for shackles in good condition. When pins become worn their breaking stress reduces rapidly. Rusting in steel shackles is also a cause of deterioration and the application of lanolin to the threads counters this in steel and makes all types easier to screw.

Shackles

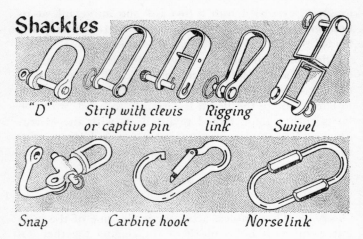

"D"   Strip with clevis   Rigging
       or captive pin   link   Swivel

Snap   Carbine hook   Norselink

If cost is an influence on your choice, bronze is more expensive than galvanised, but may have a longer life if not overstressed or abraded, while stainless is even more expensive but better from every other point of view.

Snap Shackles: These are a great help to rigging and derigging a dinghy quickly. They are more expensive than 'D' shackles and, in the case of the bronze piston types, need to be chosen with attention to their breaking loads. These go something like this.

Manganese Bronze Snap Shackles, Length		Safe Working Load	
mm	in	kg	lb
32	1¼	127	280
51	2	254	560
63	2½	356	784
76	3	508	1120

If the fitting is combined with a swivel, to eliminate twists in a spinnaker halyard, for instance, these figures are further reduced. But if the material is in stainless steel the working load may be very much greater. An example is a stainless steel swivel spinnaker halyard snap shackle, which has a claimed breaking load of 4400 lb (2000 kg).

There are other quick connectors such as the Norselink, carbine hooks and snap hooks which have applications in dinghies according to one's ingenuity and purse—for they are relatively expensive compared to shackles—and with differing breaking loads which should be considered before fitting.

Loadings on Rigging: A full investigation into this subject is beyond the scope of this book and the needs of most dinghy sailors and the only point worth making here is not to put in any weak links. The pressure applied to the rig by the wind increases by four times with a doubling in wind speed and though this is quickly dissipated by heeling or spilling wind the shock loadings can be tremendous.

So, if the designer has opted for 3 mm stainless steel wire rope rigging, of 7×7 construction and a breaking load of 1300 lb, it is advisable not to secure it to the hounds with a $\frac{1}{8}$ in shackle with a strength of only 800 lb, but to go for the $\frac{3}{16}$ in pin shackle instead.

Blocks: The simplest block consists of one sheave with a pin between two cheeks, which are elongated at one end and fitted with a bush spacer and a rivet. This end may be in the form of a metal strap, a fixed eye (either in the same plane as the sheave or at right angles to it) or a swivel eye. This block is used to change the direction of pull of a sheet lead or as part of a purchase, or tackle.

Blocks

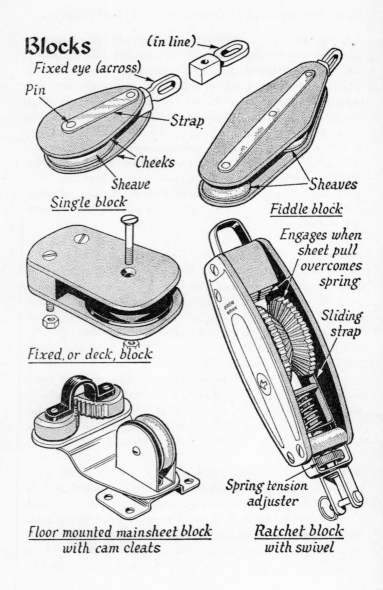

(in line)

Fixed eye (across)

Pin

Strap

Cheeks

Sheave

Single block

Sheaves

Fiddle block

Fixed, or deck, block

Engages when
sheet pull
overcomes
spring

Sliding
strap

Spring tension
adjuster

Floor mounted mainsheet block
with cam cleats

Ratchet block
with swivel

From this, there may be blocks with two or three sheaves in line (double or treble), two in tandem (the fiddle block) with the cheeks extended at both ends, a similar block with a becket (a spacer and rivet attachment point), and each with all the variations of fixed and swivel eyes. Then these are all made in various sizes to suit rope sizes and working loads. Multiple sheave blocks are used to make up purchases for mechanical advantage.

Breaking Loads: Most dinghy blocks have breaking loads far in excess of their needs and sizes are usually chosen with regard to rope sizes. With fibre rope this means choosing a block with a sheave diameter equal to, or larger than the circumference of the rope. But the best thing is to choose by inspection and to see that the rope works easily in the block without binding. If a natural-fibre rope is being used it may be wise to get a larger block than you would for a man-made fibre rope as the former tends to swell when wet. With wire rope the sheave diameter should be four times the wire circumference, or twelve times its diameter, on account of the minimum radius around which the wire should be bent.

Material Choice: Blocks are made of many materials including wood which is rather in character with the traditional type of dinghy. SRBF or Synthetic Resin Bonded Fabric laminate, commercially known as Tufnol, makes a good block for general use. The material is tough, resistant to seawater and corrosion and needs no lubrication. Cheeks and sheaves are both made from it and the latter turn well on a stainless steel pin but are sometimes bushed as well. The sides are sometimes given metal straps for additional strength.

From a manufacturing point of view SRBF has the

drawback that parts must be cut out or turned from solid block or sheet material which is wasteful and makes an expensive product for the customer. The plastics industry has provided two extensively used alternatives which can be moulded, polyacetyl resin and glass-reinforced nylon. Both have low friction, need little or no maintenance, are very tough and run on stainless steel pins with strapping and beckets of the same metal.

Stainless steel pressings are used for the cheeks of some blocks, with sheaves of nylon or acetyl. They are given lightening holes and are extremely strong and maintenance free. From a price point of view they equate with the moulded plastics and are cheaper than the SRBF but there can be large differences between manufacturers. One snag with stainless steel, which would apply to gunmetal or any metal block, is that it is more liable to give impact damage to the boat or her crew.

SPECIALISED BLOCKS

In addition to a bewildering array of single, and multi-sheave blocks, in a range of sizes, materials and end fittings, there are blocks married to jamming and restricting devices or having special features worthy of mention.

Deck, Cheek or Lead Blocks: Are screwed to the deck, the end of the boom or some inaccessible place for the purpose of reversing the direction of pull for the spinnaker leads, the clew outhaul, etc.

Snatch Block: Has an open side, a pivoting cheek or a hingeing end fitting so that a rope may be inserted into the block without the need to reeve it through from the end.

Ratchet Block: Used to reduce the holding strain on a main or a jib sheet yet becomes free to run as soon as the hand tension is partially released. It operates by the spring loaded sheave being brought into engagement with a ratchet mechanism when tension is applied, the spring disengaging it as the tension is eased.

Jamming Blocks: Kicking strap blocks usually have a jamming device integral to the design. This may be in the form of an elongated 'vee' slot into which the rope is pressed, a toothed pinion operating in a slide, which jams the rope against the sheave or an externally mounted Clamcleat.

Sheets, according to old textbooks, should never be cleated but hand held, ready to let fly in an emergency, but that was before the easily released clamcleats were designed. Modern mainsheet blocks for floor mounting may include this device to take the strain of the sheet pull. Nevertheless, in deference to old wisdom, it is always as well to ensure that the sheet slips out very smartly!

Tackle: This is an assembly of blocks and line which produces a mechanical advantage over manual effort. It may also be known as a 'purchase', when applied to a halyard tensioner, a 'hoist' when used for raising the centreplate and an 'assembly' or a 'system' describing the mainsheet and blocks.

By whatever name, there must always be a moving block or blocks, attached to the 'load' and a fixed block or blocks towards which the load can be pulled.

The mechanical advantage steps up according to how many parts (ropes) there are in the system between the moving and fixed blocks—two parts 2 : 1; three parts 3 : 1; four parts 4 : 1; etc. The rope which goes to hand

and supplies the 'effort' does not count in these ratios unless it goes directly to the load, when the advantage is increased by one.

Adding blocks into a system to gain mechanical advantage must be paid for in three ways. Firstly in the amount of rope required—a 2 : 1 system needs two feet of rope to move the load one foot, a 3 : 1 system needs three feet to move one foot, a 4 : 1 needs four to move one and so on. Secondly in the time required to operate it, and thirdly in the amount of friction which is being 'built in'. This last is particularly noticeable when the tackle is being 'run out' and not under load.

This friction is also there when effort is being applied and there could be a loss of power of 10 per cent each time a rope passes over a sheave. So, for a two-part system with a theoretical gain of 2 : 1, the power to the moving block from a manual pull of 50 lb would be only 85 lb instead of 100 lb. This is with two blocks in the system but if there were only one block and the hauling part went direct to the moving block the loss would only be 10 per cent of the initial 50 lb pull and the power to the block would be 95 lb.

So it is more efficient to pull in the direction the load is to be moved.

MAINSHEET SYSTEMS

Mainsheets are the link between helmsman and sail which controls the boom's angle, influences the setting of the sail and transmits 'feel' back to the helmsman. They are therefore worthy of study.

Transom Mainsheet System: On dinghies there may be a one-, two- or three-part purchase, with variations such as the sheet lead going to the transom or the boom, a

Tackle applied to mainsheet systems

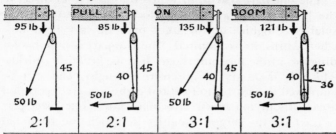

PULL	ON	BOOM	
95 lb ↓	85 lb ↓	135 lb ↓	121 lb ↓
45	40 45	40 45	40 45 36
50 lb	50 lb	50 lb	50 lb
2:1	2:1	3:1	3:1

Although we speak of mechanical advantages of 2:1, 3:1, etc. there are effort losses due to friction and bending as the rope passes each block. Large diameter sheaves reduce this

Pull direction

The Mirror's simple 2-part system (above) gives a less "downward" pull than the 2-part system below, which is combined with a full width horse

There are five effective parts to this Mirror 16 centre mainsheet arrangement

short or wide horse or a transom strop.

The one-part system, in which the sheet goes direct to the boom or via a transom block is fitted to only the smallest dinghies or to larger ones in very light airs, when sensitivity is required. The two-part is popular for dinghies in the medium range. It runs out fairly quickly, which is necessary in a sensitive dinghy, and can be employed to control up to about eighty square feet of sail area without too much muscular effort. The three-part is used on the largest dinghies in which the frictional losses and loss of sensitivity can be less important than the ability to control more sail area.

With any system there is frictional loss and the sheet comes more readily to hand if the take-off point is on the boom forward of the helmsman and then led aft to complete the system between boom end and transom.

An adjustable full-width horse allows a more central-ised setting when conditions are light, and maximum travel in the stronger winds. It enables the downward pull on the boom to be made more vertical so that a flatter sail results. If the transom block is made to travel along a wire strop it follows an elliptical path, getting nearer the transom as it is forced to the extremities. This can have a useful flattening effect on a sail during a gust.

Centre Mainsheet Systems: With the take-off point half-way along the boom and the fixed blocks amidships there is a smaller arc for the mainsheet to traverse, and if this incorporates a traveller across the boat a more vertical control can be kept on the boom over a greater sector than with the transom mainsheet. On the debit side more power is required (as the take-off point is nearer the gooseneck) which means a more complicated purchase system and the assembly takes up more room in the boat.

Four- and five-part systems are used for such cases. They are not excessively demanding of sheet length, due to the shorter arc, but do incur an amount of frictional loss. For this reason it is expedient to fit large diameter sheave blocks which have less friction.

The considerable pull exerted on the boom can give it a downward bend in the centre which removes some of the fullness from the foot of the sail. This can be reduced either by the use of a deep sectioned boom, where class rules allow, or by the separation along the boom length of the take-off blocks in the system. If the system is angled from the vertical in profile some of this pull is transmitted to the mast, bending it lower down and flattening the luff area. This is a useful control on performance dinghies.

The traveller and track come in many designs and in as many materials, with degrees of sophistication from a wire strop carrying a block, to track section in fabric laminate or aluminium, carrying a rollered traveller positioned by hauling lines. A simple system used on the Mirror 16 employs a rope strop with ends attached to the thwart about two feet apart with the lower block in the mainsheet system in the middle but not allowed to run. On the wind only the windward side of the strop tightens to flatten the sail. On most centre mainsheet systems the sheet lead can be jam-cleated in a block on the floor or the centrecase for beating. Then tacking requires only the positional adjustment of the traveller. But mainsheets should always be in hand for emergencies.

Kicking Strap: In its simplest form this can be a rope looped from the base of the mast, or to a deadeye on the floor, to a block or fixed point a little way along the boom. This performs the essential functions of preventing the boom from rising and causing a 'Chinese gybe'. A better strop consists of wire with a 'key' (to fit into a keyhole fitting on the boom) at one end and a rope purchase on the other. This can be two-, three-, and if thought necessary, four-part and should include a jamming device.

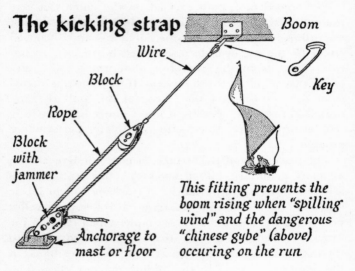

The kicking strap

Boom

Wire

Key

Block

Rope

Block with jammer

Anchorage to mast or floor

This fitting prevents the boom rising when "spilling wind" and the dangerous "chinese gybe" (above) occuring on the run

Such a kicking strap can be tightened easily and provide an extra control in the boat. A tight strap, on even a simple dinghy, will allow the mainsheet to be freed without undue lifting of the boom, so keeping the sail flat, while the forward pressure at the gooseneck can give the mast some lower bend, which flattens the sail for beating in stronger winds. With less wind and when the sail is freed off for the offwind courses the kicking strap can be eased to give more fullness to the sail.

6

Lateral Resistance and Steering

It is the sails which create the power, and the mast and rigging which transmit that power to the hull of a sailing boat, yet this would mostly be wasted, as noted in Chapter 4 (pages 71-3), but for the boat's resistance to leeway.

The ability to sail close-hauled courses, and the efficiency of reaching courses, depend on the resistance of the hull to being driven sideways and its willingness to accept the forward component of that effort. The drive from the sails, you will recall, is angled roughly at right angles to the sail chord so that when the sails are sheeted in for beating the drive is more to the side than when they are freed for reaching. For courses even further off the wind, and for running, the drive is angled forward and there is less call for lateral resistance.

The whole phenomenon is more easily understood when presented graphically and I will refer you to the 'Parallelogram of Forces' diagrams rather than attempt a detailed explanation.

THE CENTREBOARD

Unlike most keelboats, a dinghy has little lateral resistance in her hull (although displacement dinghies are better than planing dinghies in this respect) and must

Parallelogram of forces

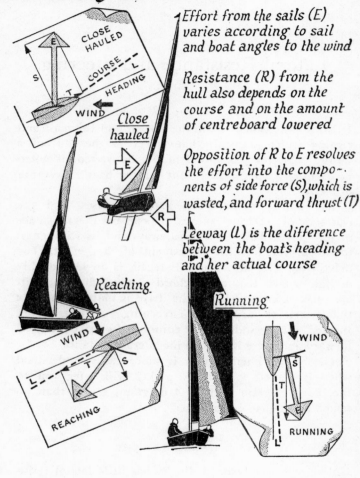

Effort from the sails (E) varies according to sail and boat angles to the wind

Resistance (R) from the hull also depends on the course and on the amount of centreboard lowered

Opposition of R to E resolves the effort into the components of side force (S), which is wasted, and forward thrust (T)

Leeway (L) is the difference between the boat's heading and her actual course

resort to dropping the flat side area of a leeboard, a daggerboard or a centre (or drop) plate. This has the triple advantages of a reduceable draft for shallow water, a reduceable wetted area when less lateral resistance is required on the offwind courses, and the ability to move the Centre of Lateral Resistance forward or aft to improve sail balance.

Wetted Area: The usefulness of the centreboard or plate is opposed by the adverse effect of its 'wetted area'. Any surface being dragged through the water causes drag, so only as much area as is required to arrest leeway should be immersed. As a rule of thumb full plate is required for close-hauled sailing, when the side pressures are greatest, 'half plate' for reaching courses and little or none for running and broad reaching. In practice, and in the light of other considerations to be outlined, this will be a subject of experiment for each boat.

Shape and Design of Centreboards: It is implicit in every well designed dinghy that the centreboard, dropplate or daggerboard will be a compromise size—big enough to stop most of the side drift yet not so big as to incur severe drag penalties. It can also be presumed to be designed to cope well with a range of wind speeds and wave conditions.

The profile shape, (whether the chosen number of square inches of side area is in a long narrow form or a short fat one) depends on the use and purpose of the boat. A short, triangular board goes well in a displacement dinghy in which the hull and its slower speed gives minimum water disturbance. There will also be directional stability gain from the long run off. A high aspect-ratio board, on the other hand, with a long leading edge, benefits the faster dinghy as the board can get

Centreboards and drop plates

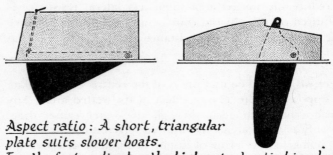

Aspect ratio : A short, triangular
plate suits slower boats.
For the faster dinghy, the high aspect ratio board can
work in deeper, less disturbed, water.

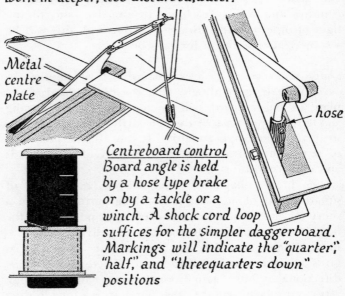

Metal
centre
plate

hose

Centreboard control
Board angle is held
by a hose type brake
or by a tackle or a
winch. A shock cord loop
suffices for the simpler daggerboard.
Markings will indicate the "quarter",
"half", and "threequarters down"
positions

down deeper into less disturbed water rather than trying to operate in that which is agitated by hull and wave interference. In most class dinghies this shape is dictated by class rules and there is little latitude for experiment but in some this may not be the case and local conditions such as sheltered inland water or a rough estuary may require a modified shape.

The shape in section and the thickness are often also matters of class rulings. In the case of wood, a thicker board is generally a stiffer one and it is most important that there should be little bending, (during close-hauled sailing when the side pressures are greatest in particular). For this reason the special 'centreboard plywood', (in which the majority of veneers are in a lengthwise direction, to give added stiffness) or a board made up of laminated $\frac{1}{4}$ in-square strips of solid mahogany, are more efficient.

The section should be shaped, to the extent that class rules allow, to give a streamlined form to lessen resistance. Both leading and trailing edges may be tapered but it is generally better to round the leading one. There is a school of thought that believes that the trailing edge should be tapered to a degree and then squared off sharply to encourage a clean breakaway. The leading edge bears the brunt of impact and grounding damage and for this reason alone should not be thinned too much. A protective strip of metal, plastics, or an inlay of bonded fabric alleviates this danger.

Less controversial than centreboard sections is the fact that the surface should be as smooth as possible to reduce surface friction. Paint systems and methods of application get attention in a later chapter, but the general aim is that the surface should be fair and smooth, but not necessarily glossy. A topside enamel with the gloss burnished off with very fine abrasive is as good as anything.

Leeway: No matter how efficient the centreboard, some leeway is inevitable on all courses. There is always the difference of a few degrees between the boat's heading and the course actually sailed—which is acceptable. Excessive leeway is not. Checking the amount of leeway is not possible when it is slight—shore transits or the line of the vanishing wake are not reliable guides as there are other confusing factors such as current and varying winds —but serious leeway is soon spotted. Included in the evidence would be the way the water leaves the transom and the slick to windward from the rudder. Comparison with other nearby boats on the same course is another sign.

The obvious cause would be insufficient exposed board but if this is not the case then it may be weed. Any slowing of the boat causes leeway, and one way of inviting trouble is by pointing the boat too high, or 'pinching'. There is an inevitable increase of leeway in stronger winds and if there is a lot of beating to do it can pay to carry less sail.

Another cause of excessive leeway is heeling. This results in a sloping board, reducing its effective depth and allowing water to slip under it. A presumed equal loss in side pressure from the sails is negated by the increased windage of the hull.

A device which is said to reduce leeway is the angled or rocking centreboard. The board is loose in its case and the width is made up by small fibre pads aft of the pivot bolt when it is vertical. Pressure on the leeward side when close-hauled causes the board to be angled in its case with the intention, if not of actually driving the boat up to windward, at least to reduce some of the

Leeway reduction

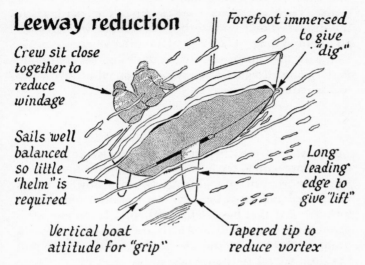

Crew sit close together to reduce windage

Forefoot immersed to give "dig"

Sails well balanced so little "helm" is required

Long leading edge to give "lift"

Vertical boat attitude for "grip"

Tapered tip to reduce vortex

Some reminders for efficient windward sailing (when leeway is greatest). Yet the most important point cannot be illustrated ~ speed through the water

leeway. Pressing a wedge into the slot at the top is a less effective way of doing the same thing but more convenient for experiment.

Before deciding that this must work remember, for a moment, what is happening when a dinghy sails to windward. Since leeway is always present the centrally mounted centreboard is already being driven at an 'angle of attack' to the course. This in itself causes drag. Will increasing this angle merely increase the drag and therefore the leeway? I must leave the question open because, although it does work on some boats it must depend on the boat, conditions and how she is sailed.

Nor does drag emanate entirely from friction, and the

side area of the centreboard which is angled to the course. There is a vortex from the tip of the board, where the disturbed and undisturbed flows meet, which causes an 'induced drag'. The effect of a similar situation can be seen with the vapour trails from the wing-tips of aircraft. Performance dinghies have tapered or elliptical boards which reduce this.

STEERING

Any boat is steered by turning the rudder at an angle to the apparent water flow due to the boat's movement. So, no movement—no steering. Given a flow of water, the angled rudder swings the stern around some point forward of it that has lateral resistance. In dinghies this pivot point of the hull is usually the centreboard but, it could be the keel or the immersed forefoot. If the hull has little lateral resistance, again—no steering.

Steering is not just a function of the rudder; when heeled a sailing dinghy will always try to luff into the wind. This is due partly to the assymetrical underwater shape of the hull in that condition, but more because the effort from the sails is outside the vertical and a pivoting movement results. Steering is also affected by the wind on the sails so that it will turn a single-sailed boat into the wind, if this is aft of the mast, and away from the wind if it is before the mast, say, a sloop-rigged boat with only a jib up. By alternately filling and emptying the sails of a sloop-rigged dinghy she may be steered without rudder.

But most steering is done by the rudder and this is efficient only when the dinghy has good 'sail balance'.

'Lee' and 'Weather Helm': For any balance you must have two opposing forces and for sail balance one is the

Steering

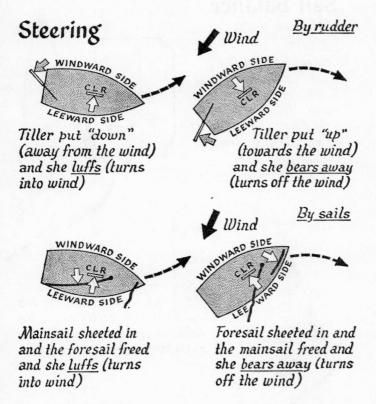

By rudder

Wind

*Tiller put "down"
(away from the wind)
and she luffs (turns
into wind)*

*Tiller put "up"
(towards the wind)
and she bears away
(turns off the wind)*

By sails

Wind

*Mainsail sheeted in
and the foresail freed
and she luffs (turns
into wind)*

*Foresail sheeted in and
the mainsail freed and
she bears away (turns
off the wind)*

'Centre of Effort' of the combined sails and the other is the 'Centre of Lateral Resistance'. (abbreviated to CE and CLR respectively). Neither point can be fixed positively as they change constantly during sailing with sail setting and the horizontal trim. The CE is the presumed point through which the force of the wind is acting on the sails and the CLR is the point through which the combined side areas of hull and centreboard could be thought to be acting. A visual idea of the latter

Sail balance

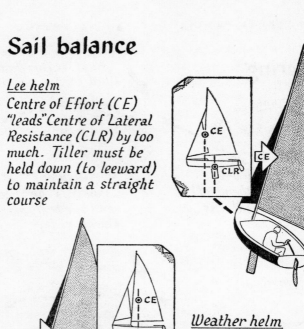

Lee helm
Centre of Effort (CE) "leads" Centre of Lateral Resistance (CLR) by too much. Tiller must be held down (to leeward) to maintain a straight course

Weather helm
CE "leads" CLR by too little. Tiller must be held up (to windward) to maintain a straight course

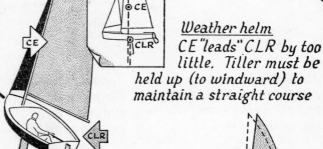

Alteration of balance
Moving the CE aft increases weather helm, (and vice versa)
Moving the CLR aft decreases weather helm (and vice versa)

is that neither bow nor stern would fall away if the hull were pushed sideways through the water at this point. The Sail Centre of Effort is more difficult to visualise, but by convention it is assumed to lie somewhere along the line between the geometric centres of the two sails. That is not all, for on most courses the sails (and the effort) are to leeward of the boat's centreline, making the thrust a 'rotating' one and tending to luff the boat into the wind.

But for the sake of simplicity the CE is usually considered in elevation only. Imagine a vertical line dropped from the CE and it will probably 'lead' (be in front of) the CLR. But for the rotating component this 'push' would cause the bow to bear away and this would be a 'lee helm' condition. If all the forces cancelled each other out the boat would be in perfect balance.

If the 'lead' of CE over CLR were less, or if it fell behind that point, the condition would apply when the 'push' would tend to luff the bow into the wind and this would be 'weather helm'.

This balance, or imbalance, has a direct effect on steering for, to sail a straight course, a continual corrective pressure must be applied to the rudder to overcome whichever imbalance you have. With weather helm, the helm (tiller for dinghies) must be held to windward and with lee helm, it must be held to leeward. Lee helm is a dangerous condition and should not be tolerated. The ideal is a slight weather helm which, when the tiller is released, allows the boat to swing gently up into wind. Close-hauled, this gives the rudder a slight 'angle of attack' to discourage leeway. But excessive helm, weather or lee, will increase drag, slow the boat and cause leeway.

Correction of Balance: Either of the two opposing factors

may be altered but the CE of the sail plan is the one with the greatest effect and must be considered first. Individual suits of sails will have different CEs and the rig may have to be readjusted to cater for this. There is a choice of canting the rig, by adjustment of the forestay and shrouds, forward or back or, if there is an adjustable mast-step, keeping the mast at the same angle but moving it bodily forward or back. Remember that moving the CE back increases weather helm and moving it forward decreases it.

The other adjustment has a smaller effect on sail balance but has the advantage that it can be done while sailing. This is the movement of the CLR forward or aft by the use of the hingeing centreboard. Raising it slightly moves the exposed part of it aft and, since this constitutes a large part of the underwater side area, this also moves the CLR and the lee helm is increased. The need for this comes on a reach when the tiller is 'heavy' with weather helm and on the beat in strong winds when there is overmuch luffing moment.

If a more permanent adjustment of this nature is required and re-designed centreboards are not acceptable by class rulings, (it may be worth seeing how yours compares with the measurements to see if there is any latitude) it may still be permissible to move the pivot hole so that, with a vertical board, the CLR is forward or back of its original position. Care must be taken, though, to see that such changes do not impede the case limitations. If there is sufficient width in the case a 'stirrup' type of adjustable pivot bolt can be used for experiment.

All this assumes that the boat has a centreboard and not a daggerboard with which practically no adjustment is possible. But in some designs the daggerboard is tapered only on one edge which means that the board can be reversed in its slot and the CLR affected in that way.

It is worth taking a little trouble to get the drop-board or plate right for the boat. It is one of the most important, and often overlooked, contributors to the performance of a sailing dinghy.

Trim: To some extent sail balance is also affected by trim, which is the fore-and-aft horizontal level of the hull and controlled, in a dinghy, largely by the weight disposition of the crew. They normally sit close together amidships for close-hauled sailing, and move slightly aft for reaching and even further aft for running. These positions are varied by wind strength and wave conditions.

Trim is adjusted partly by an instinctive or acquired appreciation of the attitude of the boat and partly by information fed to the helmsman through the tiller—the oft quoted 'feel'. Moving the crew weight aft immerses the stern sections deeper, moves the CLR aft and reduces weather helm.

This will generally be necessary when beating in strong winds when raising of the centreboard is insufficient adjustment.

Rudders: Designs of rudders, the methods of attaching them to the transom, their shape and area and all the diversifications of tillers and extension pieces are so numerous that I must just illustrate a few and write generally.

Blade design follows something of the pattern of centreboard with shallow, large blades being fitted to slower displacement boats and narrow, deep blades suiting the faster dinghies and those which sail in waves, when short blades would be uncovered in the troughs.

Aspect-ratio is even more telling on handling when the broad blade makes a much 'stronger' lever (being further away from the hinge line) and more suited to slower boats with much directional stability, and the narrow blade makes a very sensitive lever (less distance between the blade and the hinge line) and appropriate for faster dinghies.

Reduction of drag is always a desirable thing in sailing and, apart from sheer size, (no dinghy should have a larger blade than she needs) there are two main causes of drag that should be minimised. The first, surface friction and form resistance, (two linked as one) mean that the shape in section should be streamlined and that the finish should be fair and smooth. The other is to do with weather helm and leeway.

With the rudder central the difference between the heading and the actual course means that the rudder is already being dragged through the water at an angle. If it must additionally be held with its leading edge to weather at a few degrees, to counteract a strong luffing tendency, the drag is increased to an unacceptable amount. Slight weather helm, as much as can be held by finger and thumb in light to moderate winds when close-hauled, is about right and the rudder will be effectively central.

'Lifting' rudder blades, when the blade is hinged and held down by elasticated shock cord, are primarily for the convenience of fitting and unfitting the rudder assembly in shallow water, but with some arrangements it is possible to carry a partly raised rudder blade whilst sailing. This can be an advantage for reducing wetted area in light winds. As the wind increases this will give excess 'weight' to the helm on the beat and is unlikely to pay. On reaching and running courses this pays in light winds but could be risky in anything stronger—

Rudder and tiller

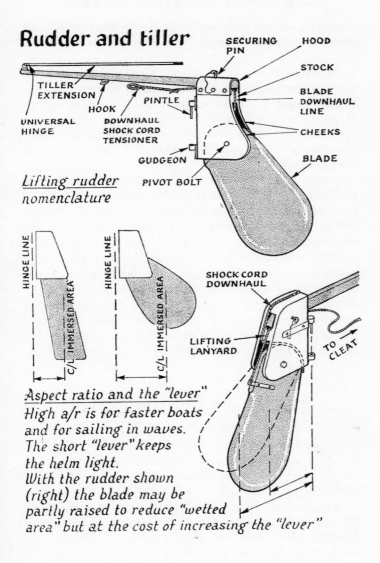

TILLER EXTENSION

UNIVERSAL HINGE

HOOK

PINTLE

DOWNHAUL SHOCK CORD TENSIONER

GUDGEON

PIVOT BOLT

SECURING PIN

HOOD

STOCK

BLADE DOWNHAUL LINE

CHEEKS

BLADE

Lifting rudder nomenclature

HINGE LINE

HINGE LINE

C/L IMMERSED AREA

C/L IMMERSED AREA

SHOCK CORD DOWNHAUL

LIFTING LANYARD

TO CLEAT

Aspect ratio and the "lever"
High a/r is for faster boats
and for sailing in waves.
The short "lever" keeps
the helm light.
With the rudder shown
(right) the blade may be
partly raised to reduce "wetted
area" but at the cost of increasing the "lever"

especially as the stern tends to rise anyway which uncovers part of the rudder.

Tillers and Extensions: These need to be suited not only for the boat but for the helmsman as well. One who is long-legged but short-armed is going to need greater length in both with possibly a crank in the tiller to clear his knees on a reach. But care must be taken to see that these do not foul the gunwales or fittings, especially during tacking.

A universal hinge for the extension is a useful but not essential item. Whatever the hinge type, play should be reduced to the minimum for here is the start of a linkage system between helmsman and rudder blade in which important feel can be lost through looseness. The fitting of tiller into rudder stock, whether by tenon or by brass hood, needs to be tight with a close-fitting pin; a smear of resin putty to build it up, or shaving down, will get this right. Then the blade must not be too sloppy in the stock and more shaving or glued-on veneer will help. But such packings must be well varnished or greased fits, and not liable to swell when wet and cause trouble whilst sailing.

HEELING AND RIGHTING LEVERS

Since the Centre of Effort of the sails is several feet above the deck and the Centre of Buoyancy, (the theoretical point through which the 'lift' from the immersed sections of the boat could be thought to act) is below the water and being restrained from leeway by the CLR, there is a considerable amount of heeling effect from the wind. The distance between these two points is known as the Heeling Lever and the amount of wind power acting on it is the Heeling Moment.

Heeling and righting

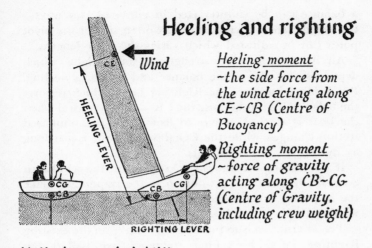

Heeling moment –the side force from the wind acting along CE–CB (Centre of Buoyancy)

Righting moment –force of gravity acting along CB–CG (Centre of Gravity, including crew weight)

Hull shape and stability

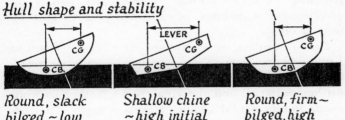

Round, slack bilged ~ low initial stability stiffens with increased heeling.

Shallow chine ~high initial stability, reduces rapidly with increased heeling

Round, firm ~ bilged, high freeboard and flare ~good range of stability

To stop the boat being blown on her side this is counteracted by the Righting Moment acting along the Righting Lever. The length of this lever is the distance between the CB, which acts as a sort of pivot, and the point representing the combined Centre of Gravity of the boat and her crew.

Without going into the subject of mechanics most people know that increasing the weight on one side of a balance can be counteracted in either of two ways—the weight can be increased on the other side or the pivot point can be adjusted which varies the lever lengths.

An increase in wind strength is the 'weight' which depresses one side of the balance and produces heeling. But heeling increases the Righting Lever which restores the balance. This is because the CB, which is central when the boat is vertical, moves to leeward as the immersed section changes, while the CG stays where it is—as long as the crew sit still, that is.

If the heeling continues the crew sit on the windward gunwale, lean out, or get out on to the trapeze, to make the Righting Lever longer by moving the CG further to windward.

The alternative is to increase the weight on the existing Righting Lever by taking on more live ballast (an additional crew member, or exchanging lightweights for heavier ones).

Hull design has an enormous effect on the Righting Lever and two factors are especially worthy of note—beam and freeboard. Beam enables the CB to be pushed further outboard and the crew to get further to windward, and freeboard enables a greater angle of heel to be tolerated without taking water aboard. For this 'flare' is an additional asset.

Application of Crew Weight: The simplest way to apply this is 'sitting out'. This can be a strenuous and precarious position so provision should be made for comfort and security. A wide side deck aids the former but if it is not available padded shorts are practical. Toestraps, if fitted, should be of strong webbing, at least 2 inches wide to avoid chafe and individually adjustable for both helmsman and crew. This can be done by buckle or block purchase. There should be enough slack to enable the backs of the thighs to rest on the gunwale. Polythene tubing must be used if the configurations of the boat do not allow straps to be draped, as this can be arched up from the floor, but adjustment is not easy. This also applies to the other leg supports such as fretted wooden extensions of the centreboard case top.

With some sort of outrigger to support the crew, weight can be got further outboard, an advantage which

Application of crew weight

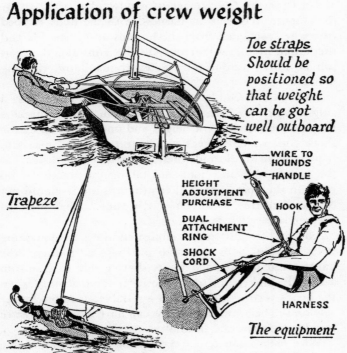

Toe straps
Should be positioned so that weight can be got well outboard

Trapeze

WIRE TO HOUNDS
HANDLE
HEIGHT ADJUSTMENT PURCHASE
HOOK
DUAL ATTACHMENT RING
SHOCK CORD
HARNESS

The equipment

Trapezing crew's fore leg is kept braced, especially on a reach, when he moves aft for purposes of trim

is reduced by the weight of the device and the amount of room it takes up in the boat. Best known is the sliding seat, fitted to the Hornet Class dinghy and used by sailing canoes for over a hundred years. Another type is the 'trampolene' or 'wing' built out from both gunwales of the International Moth and constructed of tubing, canvas and rope. Such aids provide comfort and security.

The Trapeze: Lengthening the righting lever still more and getting the CG further to windward is achieved by the trapeze. Wearing his harness, with the hook clipped to a wire suspended from the hounds on either side and the whole of his weight outside the gunwale, the crew can be fairly comfortable yet exert maximum righting moment.

Each wire assembly consists also of a grab handle, a ring to which the harness is clipped, a small rope purchase to adjust the trapeze for height and one end of an elastic shock cord which returns each wire tautly and neatly to its neighbouring shroud when not in use. The device has the benefit of being light and taking up practically no room in the boat. The skill required to use it should not be too daunting to anyone fairly nimble and with an average sense of balance.

The ring usually has an elongated alternative hooking point so that, with the same purchase adjustment, the harness can be attached to its shorter side to give him a slightly higher horizontal position for reaching, when the boat's vertical attitude is less reliable, or to the longer side on the beat, when the crew can have more confidence in a lower horizontal position.

Wire length can be checked while sitting on the side deck; a length which enables the short side of the ring to be just hooked on should be right. Remember as you ease yourself out and start to be suspended that, due to the attachment point of the wire being forward at the mast, there will be a strong swing effect towards the foredeck and the leading leg must be kept braced. Once 'on the wire' it will be possible to edge forward to the mast position to obviate this bias. On the other hand, the requirements of trim may require a further aft position, and the bracing need will be greater. Lateral balance is easily accomplished by bending the legs, while a quick return to the boat is achieved by simply slipping the feet, when one lands sitting on the side deck.

REDUCING THE HEELING MOMENT

Reducing the 'weight on the upper end of the balance' is done in three main ways: by sail control; by an alteration to the sail plan and by a reduction in area to the existing sail plan.

Spilling Wind and Sail Control: Heel can be temporarily reduced by momentarily releasing a sheet—usually the mainsheet—to 'spill wind'. Dinghies always sail best upright when there is less leeway and luffing moment and it is a sign of a good helmsman when he can do this. The 'spill' must always be of short duration or the boat will lose speed but repeated as often as necessary. Further reduction of heeling can be achieved by sail flattening or using a flatter sail, (see Chapter 4).

Lowering the Centre of Effort: The alteration of sail plan means a change to one with a lower CE, that is, to one with a lesser aspect-ratio, lesser sail area, or both. This shortens the Heeling Lever, lessens the Moment or does both at the same time. It will also reduce performance, which for dinghies is a euphemism for speed to windward, so this is not for racing dinghies sailed by

Lessening or lowering the "effort"

1. By setting smaller sails, by setting the standard sails lower or doing both

2. By reefing

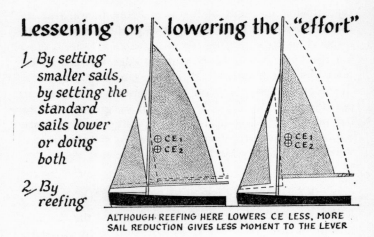

ALTHOUGH REEFING HERE LOWERS CE LESS, MORE SAIL REDUCTION GIVES LESS MOMENT TO THE LEVER

those with aspirations to winning. Yet some dinghies can have both a racing rig, with maximum sail area and a cruising rig, with reduced area. This is ideal for learners, potterers and occasionally for racing in heavy weather. A lower aspect-ratio can also be good for boats sailed mostly at sea, when the low down power helps to drive through the waves and the kinetic forces of a high mast waving wildly in the motion between peaks and troughs is reduced.

REEFING

Reducing the existing sail area by reefing cuts the heeling moment drastically. It also has an effect on sail balance, and thought must be given to that. Reefing the mainsail, for example, and maintaining a full foresail may give a degree of lee helm which may be dangerous. But it may be safer to do so in a heavy wind when full sail would be hazardous. A severely reduced mainsail, on the other hand, may be sailed without a foresail at all with less imbalance than if it were at its full.

Sail cloth does not take to being subjected to sailing stresses while part of it is rolled around a boom and the older method of using alternative cringles along the luff and leech and reef 'points' (small pieces of twine to tie the folds to the footrope or boom) is kinder to the cloth. Unfortunately few modern class dinghy sails are made with this feature.

The simplest and most common arrangement is the squared gooseneck which locates into a boom-end fitting with a similarly shaped hole. Twisting the boom, wrapping the sail around and then relocating it on to the gooseneck, holds the 'rolls' but the resulting reefed sail has the disadvantage of a covered kicking strap 'keyhole', a covered mainsheet attachment point if this is a centre mainsheet arrangement, and drooping end to the boom as the thick rolls of luff rope and linings shorten the luff more than the spiralled leech.

A boom may have tapered laths (whelps) glued to it aft, to compensate for this last trouble but there are expedients, such as rolling in sail battens, the sail bag or putting in a tapered fold in the reverse direction to the rolls.

If the reefed sail makes it impossible to attach the sheet directly to the boom, the answer is a reefing claw. This gadget, a solid aluminium casting with nylon pads to reduce sail chafe and a lower location point for strap or mainsheet attachment, holds the boom in a loose grip and needs a restraining wire to the boom end, or a rod to the gooseneck to retain its position. An alternative is a webbing strap with an eyelet in one end, which is rolled up with the sail and serves as the kicking strap boom anchorage.

These turns are put on by hand, (or hands, for it is

Mainsail reefing

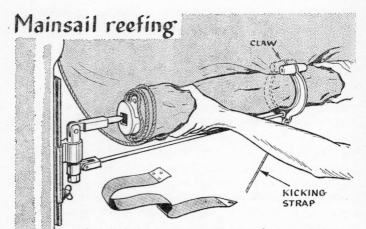

Use of the square gooseneck spigot to hold reefing turns. The loss of the kicking strap keyhole is overcome by a claw or a "rolled in" strap

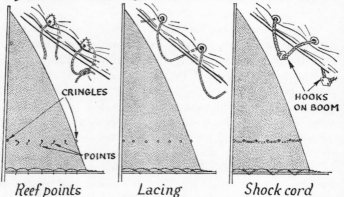

Reef points   Lacing   Shock cord

Tie down reefs ~luff and leech cringles are tied first using pennants (short lengths of cordage) and untied last when shaking out the reef

useful to have one pair keeping the luff rolls tidy while another pair pulls out the leech), but they can be put on mechanically. This is 'roller reefing' gear and consists of a wire and pulley system, a ratchet and pawl arrangement, or a worm and pinion gear.

Putting in a Reef: Partly lower the sail or hoist it partly, remove the lower batten and put on the rolls. If it is a 'tie down reef' tie first the luff and then the leech upper cringles to the boom, including some stretcher turns from the leech cringle to the end of the boom. Then, with the sail cloth between the cringles folded or bunched, tie the points under the boom, or around the footrope if the sail is loose-footed. Alternative methods of keeping this surplus sail tidy are a lacing line, which is spiralled through a row of grommets in the sail and under the boom, and a line of elastic shock cord permanently woven through a similar row of grommets in the sail and stretched down on the hooks of the boom when required.

Reefing the Foresail: Due to the normal absence of a solid spar the foresail cannot be reefed as easily as the mainsail. Yet a reduction of area is advisable, when the mainsail is reefed, in the interests of sail balance. Setting another foresail of smaller area is one way out.

The 'furling luff spar' is the only guaranteed method. The equipment consists of an aluminium spar with a groove into which the sail (specially adapted with a luff rope instead of forestay hanks) is slid. At the bottom of the spar is a drum, around which the actuating cord is wound, and drum and spar have roller bearings (at top and bottom) and are free to rotate. Furled thus, the sail is supported evenly along its luff length and may be sailed with the cord cleated and the sail reduced to any desired size.

Gunter reefing

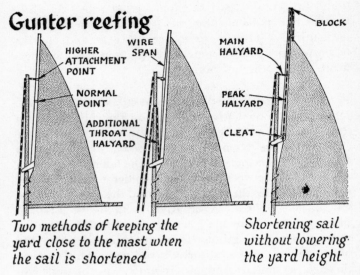

Two methods of keeping the yard close to the mast when the sail is shortened

Shortening sail without lowering the yard height

Jib Furling Gear: Similar in principle to the luff spar, with a cord actuated drum at the bottom and a swivel between the head of the sail and the halyard, this equipment furls the sail around its own luff wire.

The device needs no sail modification and is useful for the times when the foresail is conveniently dowsed. These occasions include launching and landing, especially when there are congested moorings to negotiate, and while the spinnaker is set, when a furled foresail gives a freer wind for that sail. Sailing with a partly furled sail of this type will result in the foresail unfurling at the head, giving a twisted set to the luff. For most arrangements of furling gear it will be necessary to remove the forestay temporarily.

7

Capsize and Self-Rescue

A capsize, and the skill in averting one, is all in the day's work for the racing-dinghy enthusiast; but it must be stressed here that there is a wide range of stable cruising dinghies on the market in which the less competitive sailor can embark his family with little fear of a capsize. There's all the difference in the world, for instance, between an International Moth, with its trapezes, trampolenes and toestraps, and a Dabber or Wayfarer. This chapter, however, covers questions of comfort and safety which should be of interest to *any* intending dinghy-sailor. (See page 168.)

CLOTHING FOR SAILING

The Matter of Choice: On the face of it, there are three considerations—appearance, comfort and safety; plus the inescapable factor of cost. Perhaps this last one should not weigh too heavily as good, sensible clothing is long lasting and can make all the difference in the degree of enjoyment you derive from the pastime.

Those first three considerations must be reconciled with where you sail, and in what kind of boat, for a blustery, windswept estuary, or a fast boat over which spray shoots like fusillades of arrows is going to need different attire than sheltered inland water in a non-planing dinghy. Then again, the weather, and your needs, change from season to season and from day to day so you need to have a wardrobe from which to draw.

The Neoprene 'Wetsuit': Developed for skin divers, this has had increasing use among dinghy sailors for rough water, for 'wet' boats, for boats which are liable to capsize, and for cold and inclement weather.

Waterproofing is not the main function, in fact the wetsuit is at its most efficient when water has seeped between skin and material, becomes warmed by the body, but is not allowed to evaporate. This means that a close fitting garment is required. That does not suit the dinghy sailor's need for freedom of movement, but a compromise may be a more loose-fitting one or the choice

Neoprene wet suits

A full one~piece suit and you are ready for almost anything wind and water can throw at you. For the girls there are stylish coloured suits

of one of the more flexible grades of neoprene and linings.

The material comes in several thicknesses and qualities. Among the latter are single-skin (smooth finished only on one side, which is hard to get on and off and liable to snagging); double-skin, (which is smooth both sides); nylon-lined, (which is easier to get on and off) and double-lined (which is better from all points of view). All grades are windproof, a most valuable asset in all types of sailing outer clothing.

The appearance of the double-lined material is further enhanced by colourings of red, blue and yellow, which have safety aspects. (See the paragraph on colour in this chapter.)

Style of Wetsuit: Basic choice is between the one-piece suit or the two-piece jacket and trousers. There are long- and short-sleeve, and short-, medium- and long-leg versions of these. The 'separates' have the virtue of allowing you to wear only the top or bottom half, as need and temperature dictate, but the 'short-legged and -sleeved' one-piece, which keeps the important body areas warm and gives the legs and arms plenty of freedom, has much to commend it.

The Feet and the Head: Neoprene socks are a useful accessory, even if a complete suit is not worn, and especially if your feet get wet during launching or whilst sailing. Normal sized shoes will then be too tight and a larger pair must be used. As an alternative, soled versions of the sock are obtainable.

The most popular kind of dinghy footwear is the yachting shoe with canvas uppers and a rubber 'non-slip' sole of raised hexagonal, or zig-zag tread. Low eyelet holes in the canvas make them self-draining. Ankle- or the calf-length boots may also have the non-slip sole but are

Head and footwear

A woolly hat is warm and can be covered by a roomy anorak hood for waterproofing

A cap also retains heat and is waterproof, but must be adjustable for a tight fit

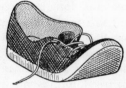

Standard canvas dinghy shoe, with rubber non-slip sole

Neoprene sock

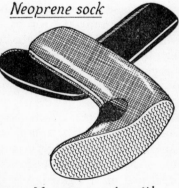

Polythene sandal

Neoprene sock, with toughened, non-slip sole and instep reinforcement

not designed to drain, which is fine, so long as wading or spray does not fill them. They are windproof and give instep protection against toestrap chafe. Yet they could be a liability during a capsize. Polythene, open-sided sandals are useful for summer wear.

The neoprene helmet which skin divers use, is too restricting for sailing. If separate headgear is preferred it is better to wear a waterproof cap or a woolly hat. The 'souwester' was popular a few years ago, until it was made redundant by the hooded anorak. Nevertheless, its virtues of being removable, not having a neck-seam to leak and giving unrestricted freedom for neck movement, may again be recognised.

Waterproof Clothing: The fully waterproof materials used are cotton, nylon or Terylene which are proofed or coated. When made up the seams are stitched, welded and sometimes taped, for strength and the elimination of leaks. A trouble with impermeability is inner condensation which is ineffectively dealt with by vent holes, better by absorbent linings, and almost entirely by the newer materials which 'breathe' out moisture without allowing it to come inward.

The lightest garments are made from the nylon cloth, with synthetic proofer. It is very strong and its lightness makes it particularly suitable for summer wear and gives freedom of movement. It is long lasting but continual crumpling may affect the proofing.

PVC (polyvinylchloride) is the coating given to the heavier garments, which may be on a base material of cotton or linen, polyester (Terylene) or nylon, or a mixture of materials. It is strong, long lasting and gives some thermal protection on a cold day, but restricts movement somewhat and does not pack so easily as the proofed nylon.

Styles of Waterproof Overclothing: Once again, there is a main division between the one-piece and the two-piece outfits. As with wet suits, the one-piece does give a little more freedom of movement but it is a case of 'all or nothing' whereas with the separate garments it is sometimes sufficient to wear the upper one.

Again, it all depends on where you sail, how fast and in what weather conditions. With habitual flying spray or the requirement to balance the boat from the trapeze or outrigger, the one-piece suit, safe against the upward-flying douche directed towards the natural gap between two-piece garments, is favourite.

Good design points include plastics zips, or Velcro, for openings with good overlaps; elasticated 'storm cuffs' at the wrists (not essential at the ankles, in fact they could trap water after immersion and make it difficult to board); reinforced seat; a hood which allows head turning without 'blinkering' its wearer, and deep pockets with drain holes and closable flaps. The position of these pockets is worth consideration. Accessibility may be limited by the design of the lifejacket or buoyancy aid which you should be wearing over the top. The 'kangaroo', or chest, pocket is a case in point and it may be better to have side pockets instead. A knee pocket is useful if you have racing instructions or a small chart to consult during sailing.

For the two-piece suit there should be a good overlap between garments. There is a choice between the 'over the head' smock and the 'front opening' styles for the upper one. The smock is more likely to be leakproof but the jacket style is more convenient to get on and off and does not disrupt hair-dos.

There is a choice between shorts and trousers for the lower garment, the latter giving much less freedom of movement. If protection is paramount, the chest-high trousers with braces, which, by the way offer more comfort and less slipping than the elasticated waistband for the active crew member, is a useful option.

Colour Choice: Waterproof overclothing is given bold, distinctive colouring—even if it is white! This has a safety factor because any sailor in distress can be spotted more easily from a distance. With that in mind, yellow is better than blue or white and red or orange is best of

Waterproof clothing

WATCH POCKET

BUILT-IN HOOD

SIDE POCKET

STORM CUFF

THIGH POCKET

ELASTICATED TROUSER BOTTOMS, PRESS STUD OR VELCRO FASTENED

Two-piece suits in pvc or proofed nylon ~ jacket and trousers; smock and shorts

One-piece suit in nylon or pvc

WRAP AROUND BOTTOMS

all. But if sailing is done inland, this has little relevance and personal preference can be given full rein.

Overclothing for a Wetsuit: The waterproof nature of neoprene means that anything worn over it is not. Instead, it is a guard against snagging, to which neoprene is vulnerable, and could also be a visibility aid, as stated above. Shorts may be sufficient for protection if balancing is done from inside the boat, with a bright lifejacket or buoyancy aid for good visibility. But for trapeze work the full suit or the high-chested trousers will give a longer life to your wetsuit.

Alternative Overclothing: For the occasions when the full protection of completely waterproof clothing can be too much, the comfort of denim, sailcloth, or one of the man-made or natural fibres, tightly woven and proofed but not given an impervious coating, can be appreciated. There is rather more water absorption if you fall in, and they take longer to dry. Styles include jackets, anoraks, trousers and shorts and a combined jacket buoyancy aid, with an inflatable collar, is available.

PERSONAL BUOYANCY

Sailing clubs, sailing schools and rescue organisations all recommend the wearing of some kind of buoyancy wear by dinghy and small-boat sailors. Cadet and school sailing associations and certain race organisers make it obligatory. Yet the main persuasion should be from the good sense of sailors themselves in recognising that, whether they can swim or not, there are dangers in the pastime, not the least of which is fatigue, which the wearing of buoyancy at least minimises.

Children especially must have this protection and it is

up to those who have them in their care to see that they are provided with the garments, and that they wear them. In mixed parties of adults and children the example should be set by the adults. I often see toddlers in sailing clubs fitted with a lifejacket, even when they are not allowed near the water and are never taken out in a boat. In this way the wearing habit is established from the start. Later, wearing the jacket will come naturally.

The Lifejacket: Many flotation aids have been graced with this name in the past and, from the appearance alone, seemed more worthy of the name than what we know as a lifejacket today, which could often be more aptly called a 'lifecollar'.

However, sailing is full of misapplied names, the important thing being to know what is meant by the accepted name, and in this case there are now very strictly applied rules as to what may be called a lifejacket.

These are laid down by the British Standards Institution from whom a pamphlet (BS 3595) on the complete requirements can be obtained. But this is unnecessary for the purchaser because the manufacturers exercise a strict quality control, and must satisfy the BSI that this is done, in order that their garments may be stamped with the well known 'kite' mark and the legend 'BSI approved'. So this is what you can look for as a guarantee that the lifejacket satisfies all the requirements.

Among the requirements are those referring to the amount of buoyancy and its disposition between the back of the neck and the front which is strapped to the chest. The total minimum amounts are 35 lb (16 kg) for the adult jacket and 20 lb (9 kg) for the child's jacket. Enclosed in a cover (also covered by a specification) and

Personal buoyancy

BUOYANCY

MOUTH INFLATION

LIFTING BECKET

WHISTLE

CO_2 CYLINDER

PERMANENT BUOYANCY

Lifejacket

LIMITS

Designed floating angle for a lifejacket wearer

AIR BUOYANCY IN DOUBLE SKIN

PERMANENT BUOYANCY PADS

Buoyancy aid

"Flotherchoc" type, buoyancy waistcoat

LIMITS

Designed floating angle for buoyancy aid wearer

strapped and secured by webbing and anchorage (which conform to a standard) the jacket is guaranteed to turn its wearer on to his back in the water (at an angle between 30 and 60 degrees) and support him high enough for breathing, even if he is unconscious.

The buoyancy may be provided either fully by mouth, fully by gas inflation, or partly by closed cell padding and partly by inflation. The gas inflation is by a cylinder which is activated automatically on immersion or manually by pulling a toggle, but there must be a visual way to see that it has been discharged. There must also be alternative mouth inflation.

When the buoyancy is partly by padding this 'inherent' buoyancy must amount to $13\frac{1}{2}$ lb for the adult jacket and 10 lb for the child's. Other regulations for lifejackets include the colouring, (either yellow or orange) and the provision of a lifting becket at the front (for being hauled out) and a whistle, attached to a lanyard and enclosed in a pocket.

A lifejacket must be worn by non-swimmers who go afloat and the one with the inherent buoyancy should be the choice, as even if the cover is punctured there will be sufficient buoyancy to keep the wearer afloat. It does cause some restriction of movement, which is a disadvantage to the dinghy sailor who must be active and duck under the boom and so on, and swimmers may prefer one of the fully inflatable types which can be worn partly inflated for sailing, leaving full till after immersion. But there is another choice, very popular with dinghy sailors:

The Buoyancy Aid: Unfortunately, this usually looks more like a jacket than does the true lifejacket—just to confuse you! It may also be in collar form or in the 'Mae West' form, and it may have inherent buoyancy, be inflatable or have a combination of the two. The essential thing to remember is that the buoyancy aid is designed to conform to another set of standards, set by the Ship and Boat Builders National Federation, and a garment so qualified will be marked 'SBBNF Approved'.

While lifejackets are mainly for offshore use the buoyancy aid is intended for small-boat users in sheltered waters, being light in weight, offering the least impediment to movement in or out of the water yet giving the wearer essential buoyancy assistance when required. The position of the floater should be vertical or angled backwards (between 0 and 50 degrees) and his mouth should be well clear of the water. A further requirement is that the garment shall be capable of repeated immersion. It should be apparent that it is easier to swim in a buoyancy aid, as the buoyancy is more 'all round' with less tendency to throw the wearer on his back.

There are more sizes and critical 'wearer weight to buoyancy minimum' ratios to be observed with these aids. Declared minimums are

18 lb   for the over 10 stone wearer
13.75 lb for the weight range 6 to 10 stone
10.25 lb for the weight range 3 to 6 stone
6.85 lb for up to 3 stone

Some manufacturers also produce garments to extend the heavy end of the range, bigger sizes and with more buoyancy.

PBAs should be chosen for comfort as well as buoyancy and a good contender in this is the proofed nylon waistcoat (sometimes fitted with a buoyant collar) marketed by Peter Storm or Flotherchoc, into which are stitched dozens of plastic sachets. Another design, the PVC waistcoat, with the linings forming a number of separate buoyancy compartments, (so that the puncture of one does not mean the loss of much buoyancy) and padded out with closed-cell foam (so that the loss is even less) can also be quite comfortable. But it should be tried for flexibility, as well as ensuring that the size and weight minimums are right.

Care of Lifejackets and Buoyancy Aids: The buoyancy garment must be in good condition to do its job, yet the nature of the pastime means that it can get hard wear without even being put to the test. The matter is worse when we see jackets and aids thrown about or being left on the floor to be stumbled over or used as cushions.

It should be obvious that safety gear must be treated with care or it will let you down when you need it most.

Keep each aid separately in a bag when not in use and see that when stored or carried it does not have to bear the weight of other things. Keep it clean, which means washing off salt and dirt with fresh water, and soap, if required, not solvents or detergents, even if it is grease. Keep the cap on the inflation valve, except when you are actually inflating or deflating, and occasionally inflate and try the valve under water to check for leaks. Check the whole appliance too, by inflating and then leaving it to see that it does not go down appreciably by the next day. Check for wear, splitting or chafe, and return it to the manufacturer for repair. In any case, return it annually for inspection and overhaul.

GETTING COLD

In our inclement climate going afloat brings a bigger risk from hypothermia than from drowning—although we must guard against both. It is more important to keep warm than to keep dry, for cold brings on rapid exhaustion with the body's attempts to replace heat losses. Its failure to do so is fatal if prolonged.

Bodies vary in this ability to keep warm in a cooling environment: children are better than adults, women better than men and fat people better than lean. But these are only generalisations and state of health comes into it as well.

Fortunately, dinghy sailors are only subjected to these conditions for a short while, after which they can go shivering into a warm clubhouse, or car, and get warmed up again. But here is a danger: we never know precisely how long we shall be out in the cold—there must always be the risk of a capsize, gear failure or the loss of wind on an ebbing tide. So it is always better to be prepared with insulating clothing for the certainty is that it is colder afloat than ashore, and the possibility of more extreme conditions arising.

But if you do get cold, leave putting away the boat or abandon her until later and get ashore, get dry with dry clothes and get warm. The last exhortation is more easily said than achieved but applied radiated heat is the most readily available, a hot shower is very good while a hot bath is best of all.

Keeping Warm: The two cooling situations are exposure to wind out of the water and in the water itself during immersion. When the first follows the second the cooling is very much more extreme.

The prevention is windproof, and preferably non-absorbent, exterior clothing and layers of insulating, air-trapping clothing underneath. The possible alternative is a wetsuit which traps air, or water, but can be dank and uncomfortable during long wearing periods. It provides no heat of itself and if you are cold to start with, or get cold, the wetsuit will not help. But it is the best clothing for immersion.

General dinghy wear is seldom bettered by the PVC or proofed nylon overclothing covering one to three sweaters and thermal underclothing. This will absorb a lot of water if you fall in and be an impediment to movement and climbing out, but even when wet, this clothing has insulating qualities in and out of the water. String underclothing is quite good but it must be topped with close fitting garments to trap the air spaces, or they will merely ventilate.

There is a great heat loss from the head, which needs covering in cooling conditions. The anorak hood helps but a woollen hat or Balaclava is better. The hands and feet should also be covered at these times, the former with dinghy gloves, but household rubber gloves with woollen glove 'linings' make cheaper, expendable alternatives. The feet need covering, not only externally with shoes or boots, but with woollen socks which still preserve some heat when wet. Neoprene socks have already been mentioned as good insulators.

BUOYANCY FOR THE BOAT

Mention was made of this in Chapter 2 but now is the time to go into the subject a little deeper. Then it was a matter of structure, crew accommodation and comfort but here the practical side must be considered.

Dinghies capsize, but that does not necessarily mean they always invert—more commonly they lie swamped on their sides. They can also be swamped by a wave or strike a submerged object and fill. In any event buoyancy is needed to save the boat and, more importantly, her crew. How much is needed? Well, unballasted wooden boats will float (although only awash) but resinglass has negative buoyancy (it is heavier than water) and a boat built of it depends entirely on the buoyancy which is 'built in'. The same applies to aluminium, while polyethylene has some buoyancy and EPS has far more than wood.

The recommendation of the SBBNF (they are concerned with a lot of other things about boats besides buoyancy aids) is the 'displacement of the boat plus 5 per cent'. With small boats, 'displacement' must also include the weight of her crew and logically, if more than the normal crew are carried the buoyancy should be increased too.

Too Much Buoyancy: A boat can have too much buoyancy but this occurs only when she has been designed with monstrous side tanks, and modern practice is to give less. The drawbacks of too much are that the capsized boat floats too high on her side, giving risk of a total inversion (from which it is difficult to recover), and that she floats high in the water (which makes it hard for a swimmer to climb on to her side). Her windage will cause problems too as she may blow away from a swimming crew or finish up on a lee shore before he can save her.

The self-draining Hull: Just as important as the amount of buoyancy is its disposition along the length and breadth of the boat. The self-draining boat has a double bottom, which can be filled with rigid foam for strength and as a precaution against loss of buoyancy through holing. The floor being above her normal waterline, she cannot lie swamped when righted and the water runs out through the transom and the centreboard slot, which is open at floor level for this purpose. But when capsized on her side she may be in an unstable condition unless buoyancy has been built into her sides as well, so countering the

Buoyancy in the boat

Enough but not too much ~ floating high on her side, her centreboard is hard to mount and she is more likely to invert or blow away

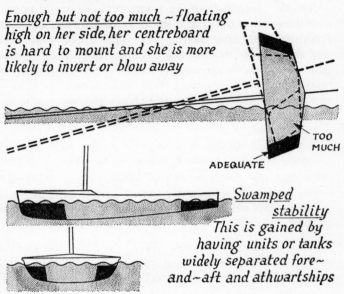

TOO MUCH

ADEQUATE

Swamped stability
This is gained by having units or tanks widely separated fore~ and~aft and athwartships

Self-draining hull
Has a double skin and a raised floor.
Side buoyancy reduces inversion tendencies when on her side

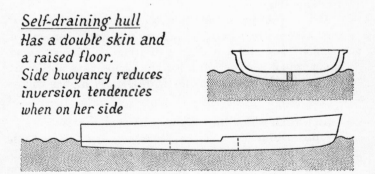

tendency to roll right over. Of course, most self-draining boats are designed with this provision.

An adverse feature of the completely self-draining hull —to some anyway—is that she is a difficult boat to board from the water after having been righted. She floats high and drifts away so that the swimmer must keep a good grip or he will lose her altogether. To experienced crews who know their righting drill this should not be a problem, for they will board as they right her, but to the less experienced, or when routines go wrong, the danger is there.

A modified version, a semi-self-draining hull, confines the floor buoyancy to the forward part so that the stern is depressed and partly swamped when righted. This cuts down drift and makes entry easier. The residue of water is then drained by other methods when the crew are aboard.

Unit Buoyancy: This takes many forms, the best known, and in many ways the best, being the PVC, inflatable buoyancy bag. Other forms include the polythene bottle, the sealed copper drum, and blocks of cork or polyurethane foam.

Accuracy is impossible when assessing amounts of buoyancy—there are variables such as the water content of the materials and the degree of inflation of containers but a rough sort of table for the supporting ability of one cubic foot goes like this, with slightly higher values in salt water.

Air—from which the weight of its container is deductable	62 lb
Polyurethane foam (closed cell)	60 lb
Cork	50 lb
Solid wood—usual small-boat hardwoods	20-30 lb
Marine plywood	25 lb

Commercial plastic buoyancy bags usually have their buoyancy value marked on them; the buoyancy value of containers can be calculated by filling them with water and weighing them. With polythene bottles this can be regarded as net, but for a metal drum double the weight of the drum must be deducted. Buoyancy tanks can be calculated by weighing the water needed to fill them—but this is tricky and usually unnecessary. Calculation can also be made by measuring volume, and if you are not a mathematical genius or do not possess a calculator I wish you luck.

The position and quantity of unit buoyancy must be arrived at by experiment and requirements. Placing them all as low as possible means less draining when the boat is righted and swamped, but this position must not conflict with the requirement for athwartships stability, for which they should be as wide apart as possible. Capsized, the low placement is unlikely to produce inverting tendencies, as boats so fitted float low anyway. The fore-and-aft swamped trim, both capsized and righted, is helped by even distribution of buoyancy along the length while the large bow bag, taking the place of a bow tank, keeps the bow up when righted and supports the considerable weight of the mast.

Security of unit buoyancy is vital. A minimum of inch-wide tape (two inches is better) must be used, with fixings of two screws and washers (or short lengths of brass or plastics strip) at each anchorage. Plastics buoyancy bags should be sufficiently, but not over-inflated, (they last longer that way) and a careful check kept on leakage. A bag that goes down during the course of a sail, or even overnight, should be replaced. A temporary repair kit is available but it is not worth risking this for long as replacements are fairly cheap.

Tank Buoyancy: References have been made to this else-where but there are a few more points to bring out. Tanks, or chambers, should be compartmented so that the loss of one by holing does not mean the loss of all. Buoyant foam can be built-in, or introduced into tanks as an additional precaution. The use of tanks for stowage, by the fitting of large access hatches, is all right if it is remembered that this does reduce the tank's efficacy and that hatch covers are less leakproof than the small drainage bungs otherwise fitted.

Class Associations have their own ways of testing tanks for acceptable amounts of leakage. For some, the test is to swamp the boat and place weights in her equivalent to crew weight. After a set period the boat is drained and then the tanks are drained separately, the quantities from these being required to be under certain limits. Other boats must be laid over on their side afloat, while crew members use their weight to depress the lower gunwale. After a period the process is repeated for the other side. Again, the boat is drained and the tank contents are measured.

A dry land method is employed by some measurers, which involves the use of specialised equipment made up of tubing, a valve, a pressure gauge and a pump. Either pressure or suction is applied to each tank in turn and the atmospheric differential must be 'held' within speci-fied limits of pressure drop (or rise) and time. A visual inspection of the tank's seams and joints is also part of the test.

CAPSIZE AND SELF-RESCUE

Nobody wants to capsize but if you are determined to exclude the possibility entirely you either do not sail light, exciting dinghies when it is windy or you sail them with well-reefed sails. Even so a tender boat may tip over in very gusty conditions.

If you own such a boat the chances are that you *will* elect to sail her in strong winds, risking the possibility of a capsize, either for the excitement or to contend seriously in racing. This is all right as long as it is not done as an act of bravado but in a full state of preparedness both mentally and bodily and with the right equipment in good order.

The mental state includes having had graduated experience in easy and then more difficult conditions, having acquired the judgement when to curtail the sail which is becoming too 'hairy' and having gained and practised the correct procedures of capsize drill, should the worst happen.

The bodily state must assume a reasonable standard of fitness and agility together with clothing adapted to the air and water temperatures, as has already been discussed, and the equipment includes well maintained personal and boat buoyancy and fittings that do not fall apart when extra strains are put upon them.

Getting Wet and Sizing Up the Situation: Although it is possible to capsize and hardly get one's feet wet (by stepping backwards on to the centreboard, righting her and stepping in again) you may as well resign yourself to getting wet. This means dropping into the water. Stepping on to the mast or the submerged centreboard case as she lies on her side is unwise as this will turn her upside down and be more difficult to right. But hold on to the boat to stop her drifting away.

Look first to the proximity of the lee shore or other dangers in that direction. The worst situation possible an uncontrolled drift could put you in is being carried under a pier, or against a groyne, a weir or rocks. Then, if the danger is imminent, the use of the anchor is the first recourse. The amount of wind is another thing to bear in mind. While it is most convenient, during recovery operations, to leave the sails up, the proximity of a lee shore or the inexperience of the crew may dictate lowering them first. And if it is very windy this may again be the best policy as the boat will be very unstable when righted, if she is not self-draining, and will probably be blown over once more before you have got aboard again.

Lowering the sails in the water is not difficult provided the halyards are not seized with jamming turns or natural fibre rope has been used for them and they have shrunk. The ultimate recourse is to undo the shackles at the heads of the sails.

Angle to the Wind: If the boat has made the usual capsize to leeward her sails will be lying downwind. Windage on the hull and the drag of the sails will reverse this, given time, so that when the boat begins to be righted the wind will get underneath the sails and flip the boat right over. There are two righting routines discussed here and the crew can prevent this happening in both cases.

Disposition of Crew: With the usual crew complement of two, the righting member (who for convenience can be the helmsman) makes his way around the stern to the centreboard and mounts it. It should be sticking right out but either he or the crew can adjust this if this is not so. When standing on this, weight should be kept as close to the slot as possible to avoid breakage. The crew can throw the helmsman the jib sheet as additional purchase for righting the boat and then take up one of two positions according to the method agreed.

For the first method he goes to the bow and grasps it.

Righting

Standard method

Helmsman climbs on to the centreboard (not the end) while crew holds bow

①

Hauls on jib sheet, or gunwale ~ boat should start to lift

②

He scrambles aboad as she rights

③

Scoop method

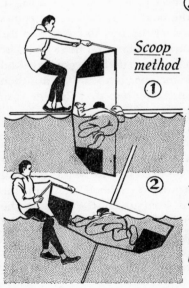

①

Helmsman hauls as before; crew gets "inside"

②

The boat rights and "scoops" the crew. The ultimate technique combines both methods

Here he acts as a drogue which tends to angle the boat head to wind—when she is righted if not before—which reduces the risk of a second flip over. With his grasp each side of the gunwale he can help to control the boat's roll and vertical attitude.

With the second method, known as the 'scoop' method, he takes up a position within the confines of the boat, lying close to the surface along the fore-and-aft centreline and holding on to a convenient projection. As the helmsman rights the boat he is in a good position to use his weight to prevent the flip over and to dampen the roll.

Righting the Boat: This is where the helmsman's agility comes in; first to climb on the centreboard, second to grasp the gunwale or jib sheet to haul the boat upright, and third to step inside the boat, if he can, as she comes up. The last feat is a matter of timing: too soon and the mast returns to the horizontal; too late and he swims. The latter eventuality does not matter if the 'scoop' method is being used, (at least one member is aboard) if not then he must board by the stern, as to do so over the side is to risk another capsize.

The Complete Inversion: To recover from this position first ensure that the mast is not stuck in the sea bed or lake bottom. If so it will be necessary to 'swim' the hull around downwind or downtide to release it. Again, the jib sheet must be brought from under the boat on the upwind or uptide side and, with the helmsman on the hull bottom, he must haul and try to depress the lee quarter, aided by the crew in the water. This should bring the mast to the horizontal when he can proceed as above.

Getting a Crew Member Aboard: With one aboard and

one in the water the questions of when and how the other member should come aboard arise. When, because if this is a non-self-draining boat it might be better to get more flotation by getting some of the water out first: how, because a dinghy is unstable and difficult to board by one in the water who is weighed down by his saturated clothing.

Some form of step is an advantage and this can be formed by a loop of the mainsheet being held by the one aboard and dangling in the water or, (for shortening of the mainsheet could be dangerous if the sail is up) a purpose-made 'stirrup' of rope, perhaps with a short piece of tubing in the loop to act as a step, and secured to a thwart.

Now, the boarding could be done over the stern, as you must board an empty boat, which is best for stability but may cause the boat to go sailing off downwind. This is caused by the drogue effect of someone in the water and hanging on. Hanging on at the side turns the boat beam on to the wind which spills much of the wind from the sails and reduces drift. Yet there is enough windage in the sails and rigging partly to balance the boarding member.

The one already aboard can use his weight not only to assist this balance but to control the lateral attitude of the boat, and to lever his companion aboard. Rather than grasping and hauling at the victim he can lean towards him to sink the gunwale on that side which makes it easy for the swimmer to get his chest and a leg well into the boat. Then, by leaning back and, if necessary, hooking his feet under the toestraps to help him, he can induce a roll which hoists the man aboard.

DRAINING THE BOAT

Earlier in the chapter various boat-buoyancy methods were discussed: after a capsize is the time when the practical value of each can be assessed.

The self-draining hull gives the crew nothing to do but get in and sail on. The first part, as stated earlier, is easier said than done but, using the boarding routines outlined and practice, little trouble should be experienced.

The semi-self-draining hull requires the crew to be aboard—with its lower flotation they should find it easier to do this—so that the boat can be 'sailed dry'. The sails are trimmed to the wind on a broad reach and crew weight is adjusted slightly to encourage the slightly 'bow up' trim given by the buoyancy. Then the combination of dynamic lift on the bottom of the hull as she moves forward and gravity flow directs the water out of the open transom.

Instead of an open transom many boats are fitted with transom scuppers, or ports, closed by shock-cord loaded flaps when not in use. This includes many boats with unit buoyancy. Once again, the technique for draining is for crew weight to be mustered aft to raise the bow while as much drive as possible is imparted by the sails on a broad reaching course.

If the boat has none of these provisions for emptying the bulk of the water then there are but two courses of action left, and both involve labour. The first is emptying by bailing. It could help to leave one crew member in the water to begin with as there is a risk, with the lower flotation caused by his weight aboard, that more water will be taken in. In any case, if the top of the centreboard case is below water level it must be stuffed with a scarf, or whatever is to hand, to prevent any flow-back. Once

Fast draining

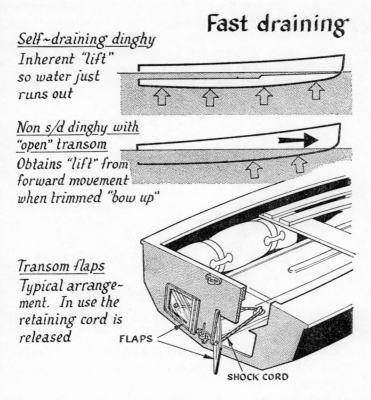

Self-draining dinghy
Inherent *"lift"*
so water just
runs out

Non s/d dinghy with
"open" transom
Obtains *"lift"* from
forward movement
when trimmed *"bow up"*

Transom flaps
Typical arrange-
ment. In use the
retaining cord is
released   FLAPS

SHOCK CORD

the level has been reduced enough he can come aboard
and help with the bailing. A large bucket is used for this
work and a 'scooping' action, lifting the water as little
as possible and chucking it to leeward, is the best one
to adopt.

The other 'ultimate' method is by beaching. A slipway is best, where she can be inched up with the transom
bungs removed. One may have little choice but to adopt
the beaching method if one lands up on a leeshore. Hope-

fully it will be smooth and flat so that, by a combination of bailing and pulling up, the boat can be emptied.

Self-bailers: A self-draining boat will sail dry but other types will probably leave a few inches to be got rid of by alternative methods. Self-bailers are the most effective

Self-bailers

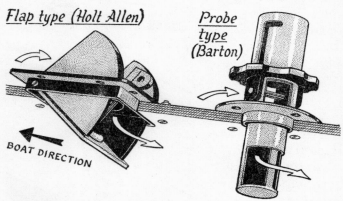

Flap type (Holt Allen)

Probe type (Barton)

BOAT DIRECTION

All self-bailers retract to form a flush skin fitting, when not in use, and require forward boat speed to operate

of these and may even work when there is a lot of water in the boat, but they need forward speed to work and that is not easy to achieve when the boat is swamped. There may also be situations when speed cannot be achieved through lack of sea-room or inability to take a profitable reaching course.

Self-bailers come in two basic designs, the probe and the flap. Both retract to form a flush fitting with the bottom of the boat when not in use and are pushed down into the water flow to achieve enough pressure differential to 'suck' water out of the boat. The tubular probe works at quite low speeds and is particularly vulnerable if left down through forgetfulness when the boat is brought ashore. The flap type is merely forced to retract when this happens.

Flap self-bailers are made in varying sizes and so can be matched to the requirements of the boat. Smaller bailers suit smaller boats, of course and require less speed, but the efficiency of any bailer is in proportion to speed which is why reaching courses are to be preferred when operating them. As speed drops so does efficiency until the point comes when back-flow occurs. Flap bailers are, to some extent, protected from this by a hingeing non-return flap while the probe type has a rubber tongue. Neither should be depended on too much.

Hand Bailers: In addition to the bucket a small bailer, with the ability to scoop as flat as possible to the floor, is necessary. The stiffer type of polythene bottle, suitably cut in the form of a scoop is adequate for this, but there is the commercial product. Also worth carrying is a large sponge which can mop up suprisingly large puddles.

Bilge Pump: This is more a feature of the larger dinghy but may have application in any. Its virtue is that it works when the boat is not moving but, in the manual sizes made for dinghies, effectiveness is low (no more than 6 or 8 gallons per minute). Yet these gallons can be discharged from where the hand bailer cannot reach— beneath a dinghy's floorboards.

A portable pump, of the plastics, 'stirrup' variety, can be kept in the locker between uses, while for the larger dinghy the brass or heavy duty polythene pump, fitted

Bailer and pumps

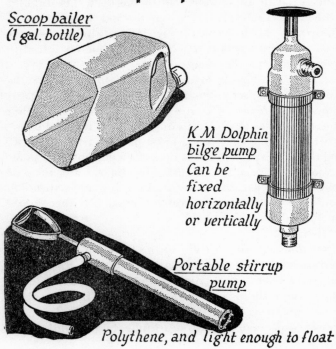

*Scoop bailer
(1 gal. bottle)*

*K M Dolphin
bilge pump
Can be
fixed
horizontally
or vertically*

*Portable stirrup
pump*

Polythene, and light enough to float

beside the centreboard case and discharging into it, is best.

RETURNING TO SHORE

After a capsize and during recovery, you will be deciding whether to go on sailing or return to shore. The answer must depend on the physical condition of the sailors, the weather conditions, how determined you are to complete a race (if you are in one), or whether the boat has suffered damage. As always, it is better to decide on the side of caution—there is always another day.

Unless it is very windy, the return to shore can be made by setting full sail and sailing carefully—but remember that numbness may have slowed the reactions of you and your crew. It may be better to reduce sail, either by reefing, or by sailing under foresail only. And then there may be gear failure, which means that you must either get home by making repairs or setting a jury rig or by calling for assistance.

Jury Rigs: If the mast has come down due to rigging failure it may be possible to set it again by the use of one of those spare shackles that the wise sailor carries in a locker. A screwdriver, a shackle key and a pair of pliers are useful to carry too, but a gadgeted yachtsman's knife can be quite accommodating. Re-stepping the mast, unless it is the shorter gunter, can normally be done only on shore.

Should this be impossible, or if the mast is broken, the boom may be held up in its place with the sail set on it. This will give a small spread of sail for a following or reaching wind to blow you along and save paddling.

A broken centreboard can make self-rescue difficult if your return is to windward (and it usually is) but if the broken piece has not been lost it may be possible, (after removing the upper half from the case and blocking the bolt holes) to wedge it in the slot. The loss or breakage of the rudder can be overcome by the use of the paddle or oar to steer with.

In fact, it should be possible to sail without a rudder at all, controlling the boat merely by the setting of the sails, but this needs a great deal of practice in light-weather conditions.

As a breed, sailors are an independent and resourceful lot and, faced with the challenge of almost any difficult situation, they tend to be able to extricate themselves and 'effect self-rescue', or just carry on sailing.

Yet determination should be tempered with good sense. If there is risk of drifting into greater danger, if there is risk to the health of the crew in prolonging the exposure, or if the damage to the boat is too great, one should swallow one's pride and call for assistance. After all, to delay in this may make rescue by others more hazardous or difficult.

Other Means of Propulsion: This could mean an outboard motor for the larger dinghy which may be able to stow it in a stern locker without too deleterious an effect on sailing. But for the average sized and small dinghy an outboard is usually a too heavy and too awkward piece of ballast to be accommodated. Ideally, such weight should be carried amidships, where its effect on sailing is less harmful, but there it is most in the way.

Paddles or oars *must* be carried. Oars are better in most circumstances, (getting along a very restricted channel could be an exception) but require more room to stow. On the floor either side of the centreboard case is the ideal, while along the side benches could be acceptable. Two paddles are better than one, so that both helmsman and crew can propel and steering is easier. Rudder, daggerboard or even spinnaker pole can be pressed into service as an emergency paddle.

Sculling: This can be done with either a paddle or an oar and is particularly useful in negotiating narrow channels. The paddle is held vertically down in the water, either over the bow, or more usually and for going backwards, over the stern. The side to side motion,

Sculling

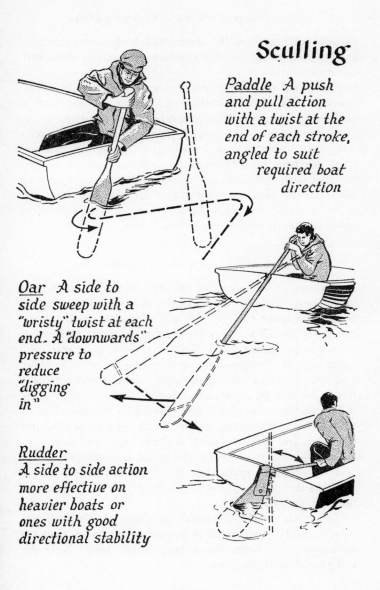

Paddle A push and pull action with a twist at the end of each stroke, angled to suit required boat direction

Oar A side to side sweep with a "wristy" twist at each end. A "downwards" pressure to reduce "digging in"

Rudder A side to side action more effective on heavier boats or ones with good directional stability without lifting from the water and with a twist at the end of each stroke, is made clear by the illustration and easily acquired.

The single oar really requires a sculling notch in the transom or a stern rowlock. There is no reason why a notched piece of wood with rudder attachments to fit on to the transom gudgeon and pintle cannot be devised. The oar (which should be straight, not spoon bladed) must this time be held at an angle down into the water. Starting with the blade horizontal and the sculler in line with the oar and grasping the loom with his wrist, or wrists, vertical, a shallow 'figure of eight' with a twist at the end of each stroke is followed.

CALLING FOR ASSISTANCE

If you race or belong to a club and sailing is well organised, there will be a manned rescue boat either in attendance or on call. This is why dinghies which call on voluntary rescue services are seldom in these categories. If you sail unchaperoned then, obviously, your responsibility will be greater—not only to be more self-reliant and judicious in choosing the conditions in which to sail, but to be able to attract attention.

Signals must be visual and waving is the simplest of these. This is where the visible 'safety colours' of clothing come in and the accepted 'distress wave' is slowly to raise both arms from horizontal at the sides to the vertical and lower them again, but this may not be easy in a pitching dinghy so taking off a smock and waving it may be more practicable.

'Sail in the water' is another (though unofficial) signal which could be employed if it were necessary to attract a helicopter, to whom this is understood. Most likely the sail would be there anyway.

For inshore coastal waters flares should be carried and a small pack of three, designed for dinghies and sealed in a polythene wrapper, is available. These hand 'fireworks' are usually red magnesium flares or orange smoke, the latter being better for attracting searching aircraft. The operating instructions should be closely followed, especially the one that directs that the flare should be held at arm's length and pointed downwind.

Accepting a Tow: Do not assume that the approaching rescue boat is prepared to offer a tow. The main concern with any rescue is for life, and the rescue boat may neither have the ability, nor her owner the wish to undertake any assistance for the boat. For your part, be ready to accept rescue on these terms and to abandon the boat. Drop the anchor or let her finish up on a lee shore and return for her later.

For the tow, be sure that the boat is upright and with as little water in her as possible, that the sails are lowered and tidy, that there are no trailing ropes, and that if the mast is down it is lashed to the deck with the minimum protruding at the bow. If both crew can leave the dinghy and go aboard the rescue boat it will lighten her load; but if one stays behind he can prevent her yawing by steering while bailing with one hand.

If your painter is strong enough and attached to a through-bolted cleat or the mast, this can be used; otherwise the rescue boat may provide the tow-rope—which must be attached to a strong point. The length of this stern tow-rope is best kept short, this way it reduces yawing.

An alternative way of towing is for the two boats to be lashed side by side, with fenders between. A rescue boat may prefer to do this, particularly if it is necessary to thread a way through moored boats as a tighter 'rein' is kept on the towed boat and the steersman can keep his eyes to the front without wondering what the tow is doing.

Thanks: The rescued may be too numb to thank their saviours right away but the act should not be forgotten. A little monetary reward should also be pressed, perhaps to go to club funds if this is their rescue boat, or the crew could be treated to refreshment if the hostelry is near. One hears of salvage claims being made by rescuers in respect of larger craft but, although still legally tenable, I have heard of no instances in respect of dinghies.

8

Wind, Water and Helmsmanship

WIND AND WEATHER

The Lengthening Season: Time was when the sailing season started in May and finished in September, but in latter years it has been starting earlier and finishing later —or not finishing at all. This is no thanks to the weather or to a hardier or more enthusiastic breed of sailors, but to better thermally insulated clothing and to boat materials like resinglass and polyester which are more tolerant of wintertime abuses. The weather, alas, shows no change for the better.

Yet in fact there is very often better sailing weather out of the normal sailing season than there is in it, while sailors who ritually delay their first launch until April can run slap into some of the most boisterous conditions of the year.

Some of the advantages of the winter months for sailing are that winds are usually 'truer' and there is less difference, than in summer, between the wind speed in a gust and its succeeding lull. Others are that, for inland sailing, the lack of leaves on the trees make for less wind deflection and thermal influences and, of course, there are fewer boats around 'cluttering up the waterways'.

This is not to advocate winter sailing over summer sailing, but to emphasise that good sailing weather should be recognised when it occurs.

Wind, our Meteorological 'Consumable': Sailing makes

the study of the wind, and its place in the pattern of the weather, particularly rewarding and if you wish to probe deeper into the subject the book, *Wind and Sailing Boats* by Alan Watts (Adlard Coles Ltd) can be recommended.

Wind comes firstly from planetary air streams, due to the rotating of the earth, (of which the temperate Westerlies is the one which affects our weather); then by pressure systems within that stream, (depressions, anticyclones and so on); and then by the more local phenomena of day and night temperatures and sea and land effects.

Weather forecasts, given on radio and television and in the newspapers, tell us what to expect, and from synoptic charts and from the behaviour of a barometer and the appearance of the sky it is possible to predict a little more.

Some forecasters and most sailors use the Beaufort Wind Scale which divides wind velocity into 12 Force numbers each with a range definable in knots.

Beaufort

No	Description	Knots	Dinghy performance
0	Calm		
1	Light air	1 to 3	No sitting out
2	Light breeze	4 to 6	Helmsman may sit out
3	Gentle breeze	7 to 10	Helmsman and crew sit out
4	Moderate breeze	11 to 16	Some planing
5	Fresh breeze	17 to 21	Some spilling of wind on the beat
6	Strong breeze	22 to 27	A dinghy should reef
7	Near gale	28 to 33	Survival sailing
8	Gale	34 to 40	None

It should be remembered that predicted Force numbers are averages, and that velocities can exceed these in the gusts. I record only the lower end of the scale here as a dinghy is sailable only up to about Force 7.

With the reference to reefing it should be remembered that wind pressure on the sails increases as the square of the wind speed, so that an increase of twice the speed creates a pressure four times as much.

The main trends of the weather are given on the synoptic charts. Among the points to note are that depressions in the pressure system result in an anti-clockwise rotation of the air 'filling them in' and that anti-cyclones result in a clockwise 'spill off'. These are 'gradient winds' which roughly follow the contour lines of the isobars (lines which join points of equal barometric pressure), but with a bias towards the pressure adjustment direction. The greatest pressure differences, which show in the charts by the isobars being closer together, produce the steepest gradients and therefore the strongest winds. Widely spaced isobars give calm conditions.

Another point to note is whether the pressure systems are in motion (usually tracking eastwards in the planetary air stream) or stationary. If the former is the case the speed of the movement is worth noting, and whether the centre, of, say, a depression, is passing to the north or the south of you. If to the north, the wind veers (changes direction in the same way as the sun, *e.g.* from SW to NW) as the system moves from west to east. Conversely, if the centre is to the south of you the wind will back (the opposite way to the sun, *e.g.* from SE to NE).

How far this tracking has progressed, and just where the centre of a depression is, can be roughly established by standing with your back to the wind; the centre will then be on your left. How long it takes for a pressure system to pass over depends on its speed, and can vary

Chart of typical depression

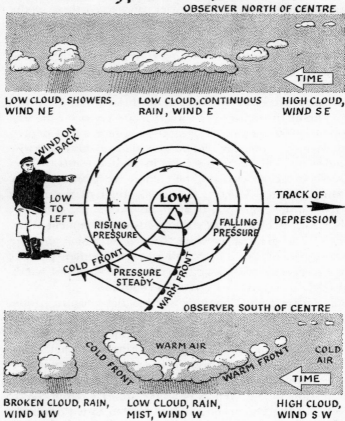

OBSERVER NORTH OF CENTRE

TIME

LOW CLOUD, SHOWERS, WIND N E   LOW CLOUD, CONTINUOUS RAIN, WIND E   HIGH CLOUD, WIND S E

WIND ON BACK

LOW TO LEFT

RISING PRESSURE

COLD FRONT

PRESSURE STEADY

WARM FRONT

LOW

FALLING PRESSURE

TRACK OF DEPRESSION

OBSERVER SOUTH OF CENTRE

COLD FRONT   WARM AIR   WARM FRONT   COLD AIR

TIME

BROKEN CLOUD, RAIN, WIND N W   LOW CLOUD, RAIN, MIST, WIND W   HIGH CLOUD, WIND S W

*Note: Main differences between being north or south of
the centre are~ (a) wind backs to the north, veers
to the south (b) absence of fronts to the north,
changes are smoother and less vigorous than south*

from a few hours to days or weeks. The dinghy sailor can take an interest in these systems to get a background picture of the weather and what winds he can expect, but the short duration of his sailing means that this expectation may be very much influenced by local conditions.

Friction Layer: Isobar pressures are taken at a height of 2000 feet, which means that these pressure systems are at this height too. We sail in a layer of air called the 'friction layer' close to the ground, or water, and what this air is doing may be very different from what is going on up above. Convection, or vertical movement, is occurring all the time in this layer, caused by temperature variations on the surface. These are due to the heat absorption differences between night and day, sea and land, forest and cornfield, town and country and many others. Air moves slower at the surface (due to friction) so the descending air that replaces rising air will be faster moving (gusts) and may also be from a different direction. Clouds too, especially the low, cumulous ones, carry cool shadows before them, bringing down gusts from above.

Friction is less over water than over land and this can affect not only the speed of the wind (mean speeds are usually twice as great over the sea as over the land) but its direction as well. This occurs as the wind passes over the coast, where there is an apparent 'bend', and also over lakes and rivers, where a shallow 'Z' bend occurs. The physical contours of the land also affect both wind speed and direction, having channeling and blocking effects, while trees and buildings do the same thing in more localised areas.

Thermal Wind: This local wind generates as the land begins to warm up in the morning, bringing in cooler air to replace it from over the sea and reaching its maximum

Veering gusts and friction effect

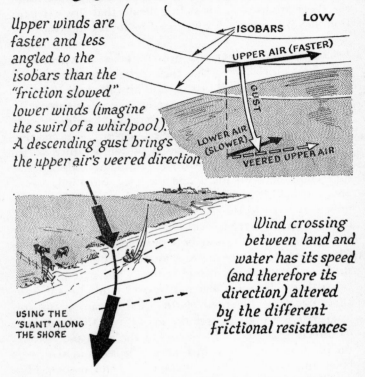

Upper winds are faster and less angled to the isobars than the "friction slowed" lower winds (imagine the swirl of a whirlpool). A descending gust brings the upper air's veered direction

ISOBARS

LOW

UPPER AIR (FASTER)

GUST

LOWER AIR (SLOWER)

VEERED UPPER AIR

USING THE "SLANT" ALONG THE SHORE

Wind crossing between land and water has its speed (and therefore its direction) altered by the different frictional resistances

strength in the mid-afternoon after which it begins to decline. This is the 'sea breeze'. It is reversed at night, when the sea is warmer than the land, by the 'land breeze' but this is of no great interest to most dinghy sailors. The sea breeze may be opposed to the prevailing gradient breeze, may be blowing the same way or may blow at any angle to it. By blowing the same way it augments the gradient breeze, and can easily turn a manageable Force 4 into an unsailable Force 6. Opposing the gradient wind, it becomes a trial of strength which can result in variable winds or even a complete calm, but is most likely to give a sea breeze in the afternoon when it is at full strength. With the pressure gradient wind at any other than these two directions the sea breeze will augment or reduce, according to the relative angle at which they meet. The resultant wind direction depends on the vectors of the two directions and strengths.

The sea breeze develops most strongly in clear sunny weather, which makes it the despair of seaside sunlovers! If there is damp air over the sea, however, this is drawn in and condenses into cloud when convected over the coast. This reduces the thermal efficiency of the land and the sea breeze may stop.

Thermal winds occur inland along river banks and around the shores of lakes, especially when they are bordered with good thermal sources such as arable land, but these are of relatively low intensity and are most noticeable in very calm weather.

Gusts: Wind in the friction layer is constant neither in speed nor in direction. Gusts are usually $1\frac{1}{2}$ times the mean wind speed which means that a Force 4 could gust to Force 6 while the lulls could be Force 3 or less. The dinghy sailor has not only to balance his boat through this range but must remember the sailor's maxim—'reef for the gusts, not the mean wind speed'.

The directional change is caused by the faster moving upper air that is brought down to the surface (the constituent of the gust) being veered to a greater extent than the slower moving air. I hope that the diagram makes this clear.

If not, it must be taken on trust that in most cases, a gust will veer from the true wind direction. From the

Convected wind

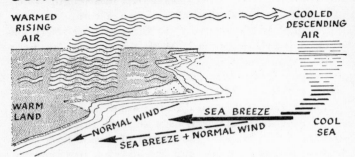

WARMED RISING AIR

COOLED DESCENDING AIR

WARM LAND

COOL SEA

NORMAL WIND

SEA BREEZE

SEA BREEZE + NORMAL WIND

Sea breeze This is a flow of cooler air from the sea, replacing the rising, warm air over the land and completing a cycle. If a normal gradient wind is also blowing a vector is formed

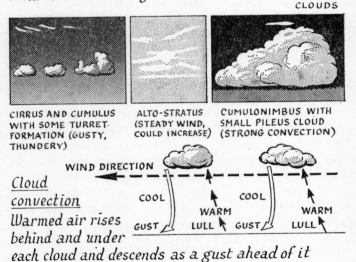

CLOUDS

CIRRUS AND CUMULUS WITH SOME TURRET-FORMATION (GUSTY, THUNDERY)

ALTO-STRATUS (STEADY WIND, COULD INCREASE)

CUMULONIMBUS WITH SMALL PILEUS CLOUD (STRONG CONVECTION)

WIND DIRECTION

Cloud convection

Warmed air rises behind and under each cloud and descends as a gust ahead of it

COOL

WARM

GUST

LULL

COOL

WARM

GUST

LULL

sailor's point of view this is of advantage if he is on starboard close-hauled tack, as the freeing of the wind gives him a 'lift' to windward. On port tack the wind 'heads' so that he must either bear away or tack.

The Influence of Clouds: The influence of clouds on the friction layer in which we sail depends on their height. The feathery cirrus clouds at a great height may presage strong winds and the approach of a depression, but by the time it arrives the dinghy sailor will have had his sail, packed up and gone home. Layer cloud, alto-stratus, of lesser height, may be the advance of the depression but the winds should not be 'fluky' although they may gradually get stronger and your sail may have to be curtailed. It depends how steady the barometric pressure is; if it is about to plummet this should have been forecast.

Cumulonimbus is the thundercloud. It rises to a great height but its base is low and it creates tremendous up-draughts in its centre and strong downdraughts around its edges. Obviously the resulting squalls are going to put the dinghy sailor in trouble and he would be wiser to stay ashore.

Cumulus, the puffy, fair-weather clouds, are at low height and show evidence of thermal activity. As the cold 'shadow' of each cloud falls across the dinghy sailor he will feel the gust of its downdraught, followed by the lull of the rising air in its wake. These clouds are often in lines called 'streets', stretching along the direction of the wind. Between the cloud lines the convection currents will be less strong than under them.

Apparent wind: Put your hand out of a car window on a calm day and you will feel a wind equal to the speed of the car. This is 'apparent wind' and the same thing happens in a boat—with the differences that a sailing dinghy cannot sail in a calm nor directly into the wind. But she can sail directly downwind and if her speed is 4 knots 'away' from a true wind speed of 12 knots then the apparent wind speed is 8 knots. Beating into the

Apparent wind

Vector diagrams give both speed and direction. An increase in wind speed enables the boat to sail "freer". A faster boat speed "heads" the wind so she must sail closer

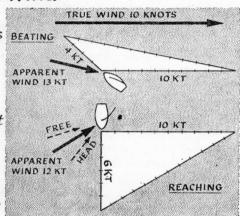

wind the apparent wind speed is logically increased but since your movement must be at an angle no less than 45 degrees from the true wind you must draw a vector diagram, with the relative speeds to scale, and measure the resultant if you want to know the apparent wind speed. This diagram also shows the apparent wind direction.

Vector diagrams can be drawn for all courses and for the countless permutations of wind and boat speeds. It is worth noting that the burgee lines up with the apparent wind.

Relative or Surface Wind: The sailing dinghy, as has just been noted, cannot sail in a calm, for it feels no wind, but if the water on which it sits is itself moving in a current then it is in the same condition as the car which is being driven through still air. Your sails feel a wind—the relative or 'surface wind'.

A current or tidal flow of 5 or 6 knots will produce a surface wind of that amount which, when the wind is opposing the current, can make a substantial increase to the wind felt in a dinghy. Similarly, if the wind and the current are blowing the same way, the wind speed must be reduced by the speed of the current. If the two flows are neither directly opposed nor coincidental, but at an angle to each other, a vector results showing either a proportional increase or decrease, together with a change in the apparent direction of this 'new' wind.

On a river this can result in a host of apparent wind directions as the river bends, while on the sea or estuary the change of the tide direction can make a vast difference to the wind and sea conditions. A 'wind over tide' situation may give quiet sailing but when the tide changes and it becomes 'wind against tide' both wind and sea conditions become noticeably rougher.

CURRENT AND TIDAL FLOW

Having got used to the idea of vector diagrams showing how wind is influenced in speed and direction, it is almost stating the obvious to say that the boat's own progress is affected, both in speed and direction, by the movement of the water she is sailing on. Yet it can easily be forgotten and may give the sailors a longer trip than they bargained for.

This is particularly so for the slower dinghy, which may have a speed of only 3 knots or so, while currents often run at 4 knots or more. Plugging into the current the dinghy's speed (over the ground) is reduced by the speed of the current while, conversely, when travelling with the current her speed is increased by that amount. At an angle to the current (which she will have to do when tacking into it) the vector diagram may show that her 'gain' over the ground is negligible.

Surface wind relative to current

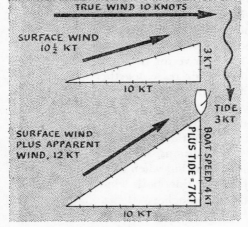

To calculate speed of wind directly "over" or "against" current, simply add or deduct. If they are angled to each other a vector diagram can compute both speed and direction

Sailing in tidal areas it is thus important to gain local knowledge of the tidal flow with its special characteristics, hazards and patterns. Lateral direction, channels and counter-flows, the times of tide change and its height should all be assimilated.

General tidal information is best obtained from a yachting diary, an almanac or a tide-table which can be obtained from chandlers. If you get the local one it will have the exact prediction for that area, and save you working out the time allowances from high water at Dover, say. There may also be a list outside the harbour-master's office. The newspapers publish day-to-day tidal details but it is better to have the predictions for the year as it is then possible, by examining the trends, to tell whether you have Spring or neap tides to deal with. The greater height of the former may not be of importance but the fact that this tide will produce faster tidal streams will be.

The safest thing to do is stay 'up-tide' which is better than staying up-wind. Should the wind drop or gear fail it is then easier to get back to the launching place. But if the tide is due to turn this does not apply as you can go downtide with one stream and you are then upstream when the tide turns.

'Cheating' Tidal Flow and Current: Whether cruising or racing the time comes when you must go against the tide and, if you want to make good progress, the thing to do is to sail in the weaker-flowing streams. Picking these is a combination of knowledge and observation.

Water flows fast in the deep channels and slow, due to surface friction, over the shallows. Places to avoid are the main channels of estuaries and rivers: if these are not marked by buoyage, the fact that boats do not usually moor in the fairway may give an indication, as will the slight decrease in your boat's progress as she stirs mud. Appearance of the water may also give a clue, showing either rougher water as the stronger 'wind against tide' situation occurs, or smoother water, where wind and current flow in the same direction. A chart of the area will also yield much information.

In a river the same situation obtains, and the current will be slacker in the shallows, which should be near the banks, but not around the outskirts of bends, which will often have the deepest water close to that side. The inside of bends should not be hugged too close, as silting there may have you aground.

Close observation may show evidence of a backflow

(against the flow of the mainstream) which can be taken advantage of. Estuaries often provide examples where the tide turns sooner in one channel than another or one side of the estuary has changed its tidal flow, while the other side is still using up the remains of the old tide. Inlets and harbours have their own circulations and a harbour mole, around the end of which the tide may be sweeping, may have a strong counter-current on its 'downtide side' which may provide the easy way of passing it.

Evidence of counter-currents is given by the headings of moored boats, although one needs to be fairly certain that they are not merely being 'wind-rode'.

'Lee-bowing': When sailing close-hauled into an adverse current, with the wind, slightly angled to that stream, one should be able to 'lee-bow' on one tack. When the bows are being headed directly into, or slightly to windward of the current, pressure on the leeward bow pushes the boat crabwise to windward. It is worth 'pinching' a little to obtain this effect, which makes this a very rewarding tack.

Moored Boats and Current: Sailing around and between moored boats requires care if collisions are to be avoided. Always pass downtide of them. As most moored boats are moored only at the bow, and all swing together when the tide changes, passing downstream usually means passing under their sterns. The temptation to 'weather' the bow must be resisted.

Launching and Landing in Current: For a normal launch with an onshore wind you start off with the tack that takes you offshore quicker, but the choice may be influenced by the effect of current. It is generally wiser to start by heading into the current. This way things happen

Current and waves

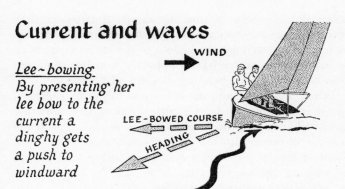

Lee~bowing
By presenting her lee bow to the current a dinghy gets a push to windward

Course A ~ sets boat downtide

Course B ~ straight, by heading her uptide slightly

Course C ~ uptide to start with in the weaker current.

C is safest and often fastest

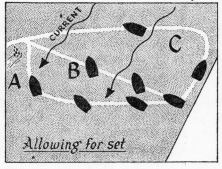

Allowing for set

Wave motion
Water particles have an orbital motion as each wave passes. The forward thrust of the crest plus gravity can give surfing, or "wave planing"

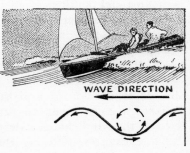

WAVE DIRECTION

more slowly, there is time to avoid obstructions like moored boats and you are under better control. Additionally, you may well get the benefit of the lee-bowing effect, so that the tack which has an apparent heading along the shoreline has a true heading directly offshore.

When making a landing, it is again better to approach the shore against the lateral current, using it to slow the boat and keep her under control. This is fairly easy if the wind is blowing with the current—you merely luff, spill wind or harden the sheets as required—but if the wind is blowing against the current it may be necessary to reduce sail to control your speed.

The exception to making a landing against the current is when the sails are down and you are making the approach under oars or paddle.

WAVES

These water undulations, which we must face if we sail on open stretches of salt water or even on some fresh water lakes and reservoirs, have great stopping power on a light dinghy, but can also give her great impetus when conditions are favourable. They also cause difficulties at launching and landing.

Waves are produced by the friction of wind on the surface of water, and wave height depends on wind strength, on 'fetch', (the distance over water the wind has travelled) and on time during which the wind has been blowing. But these are 'textbook' waves which are seldom allowed to develop; and if they do dinghy sailors are unlikely to meet them. In practice there is insufficient fetch or the wind changes, (leading to a confused sea) or the tide changes, (altering the 'effective' wind speed). This latter is the 'wind over tide' or 'wind against tide' situation which can give a sudden change in the sea state.

There are also the 'residue' waves—the dying ripples of storm centres which may have occurred days ago and hundreds of miles away.

The speed of a wave may be around 8 knots, by which is meant the impulse and not the actual water in that wave. Instead, the molecules of water describe a rotary motion, returning to the same point in space or perhaps a little further forward, as each wave passes. Watching a floating cork illustrates this. But on the crest and the back of the wave the motion is forward, while in the trough the motion is backward.

Sailing in Waves: When beating into waves the aim is to lessen the impact of the forward moving parts of the wave by driving the bows directly into it (rather than letting the wave stop the boat by striking the flat side of the bow) and getting the crew weight, (and any internal ballast) centred amidships, to encourage the bows to lift and the boat to 'ride' the waves. Over the crest a slight bearing away will maintain speed. Of course, a heavier dinghy may need to be sailed rather differently, will have the weight to allow being driven into waves, and will require less bearing away to maintain speed. But no boat will 'point' so high in waves as in calm, and all will gain from being driven slightly freer.

The Wave Plane: This is really surfing with the wave and is a gravity pull as the crest catches up and lifts the stern. For a while, and encouraged by the crew moving their weight forward and then back to balance, the slide down the forward face of the wave can be balanced by its lift, and the boat goes at wave velocity. This is most usual on a run or a broad reach but conditions may also be found right on a beam reach, when planing across the waves.

Launching and Landing in Waves: As waves approach the land they 'feel' the shallowing sea bed by the bottom half of their cycloidal motion being cut off, the upper, forward part carrying on as surf. A gently shelving beach gives diminishing wave height and there is little difficulty for small boats—unless it is rough, when they should not go out anyway.

A 'steep-to' beach can provide difficult surf in moderate conditions when the sea state further out is amenable. This is so often accompanied by an onshore wind that only this case is worth considering here.

This enables full sail to be set on the shore before launching, as the boat will be positioned and kept head to wind by her crew. It should be stressed here that there is no power in the sails while they are allowed to flog and that this condition is the only safe one for raising and lowering sails without snagging them, for keeping control of the boat at the start of the launch and for carrying out sail adjustments afloat.

It is worth considering, too, whether or not to fit the rudder on the shore. Many rudders have been broken by a wave picking up the boat and tossing her back on her transom. With the crew on one side in the water and holding the boat and the helmsman on the other the boat is pointed directly into the waves, walked out gradually and allowed to 'see-saw'. Choosing a lull in wave height first the crew jumps aboard then, with a shove, the helmsman does likewise. Paddles must then be used to get beyond the surf (the rudder can be used as one paddle). Sail control may be used by the crew, (pulling in the mainsheet to luff; releasing it or pulling in the foresail to bear away, with a little centreboard down) if the helmsman has to fit the rudder. (Illustration on p. 131.)

When landing in the surf of a steep-to beach it is best to lower the mainsail when there are only a few dozen yards to go. The centreboard can then be raised most of the way and the helmsman can loosen the pins ready for the rudder to be removed. At the last moment the rudder is removed and brought into the boat, the crew puts a leg over the side to feel for the bottom and, at the moment of grounding, slips over the side with the helmsman to run the boat up the beach on a wave. During the approach and until clear of the waves, the boat must be kept stern-on to the waves.

SPECIAL SAILING TECHNIQUES

Sailing dinghies need to be balanced by crew weight at all times, both laterally and fore-and-aft, and this was stressed in Chapter 6. This is particularly so in the strong and light extremes of wind conditions. There are also adjustments that can be made to the boat and her rigging to assist at these times, and those applying to the sails and rigging were discussed in Chapter 4. But there are some further points about handling:

Beating in a Strong Wind: In addition to flattening the mainsail by adjustment it may pay to have a flatter-cut foresail for these conditions. This must be set before going out and if there are more running and reaching courses than beating this may not pay. The mainsail sheet traveller can also be set to its widest stops which will give poorer pointing but develop more power.

With so much drag from the wind it is essential to keep the boat moving well. This is achieved by a combination of bearing away slightly in the gusts and spilling a little wind quickly from the mainsail, if necessary, to reduce wind pressure. Endeavour to keep the minimum of heel as too much accentuates a 'screwing into the wind' tendency, which stops the boat. Luffing, as a means of spilling

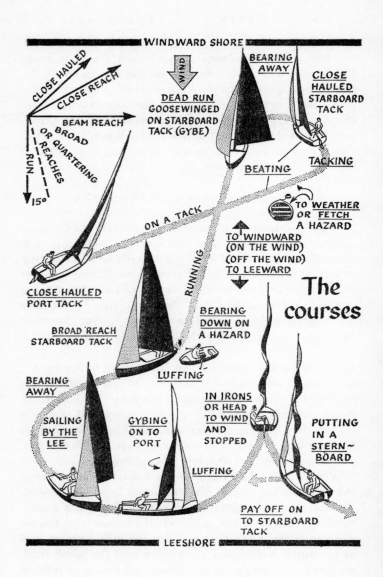

WINDWARD SHORE

CLOSE HAULED

CLOSE REACH

BEAM REACH

BROAD OR QUARTERING REACHES

RUN

15°

BEARING AWAY

CLOSE HAULED STARBOARD TACK

DEAD RUN GOOSEWINGED ON STARBOARD TACK (GYBE)

WIND

BEATING

TACKING

ON A TACK

TO WEATHER OR FETCH A HAZARD

TO WINDWARD (ON THE WIND) (OFF THE WIND) TO LEEWARD

RUNNING

CLOSE HAULED PORT TACK

BROAD REACH STARBOARD TACK

BEARING DOWN ON A HAZARD

The courses

LUFFING

BEARING AWAY

SAILING BY THE LEE

GYBING ON TO PORT

IN IRONS OR HEAD TO WIND AND STOPPED

PUTTING IN A STERN~BOARD

LUFFING

PAY OFF ON TO STARBOARD TACK

LEESHORE

wind may be good practice in the case of a heavy dinghy, with the weight to carry her through, but not for the light one. Crew weight should move further aft as the wind gets stronger.

Running in Strong Winds: For the dead run, sailing goosewinged (mainsail and foresail on opposite sides) results in a steadier boat. Rolling can be further reduced by lowering the centreboard and by the helmsman's control of mainsheet and tiller—drawing in the sail slightly towards the roll or steering towards the roll.

A dangerous condition is 'sailing by the lee' when the burgee flies on the opposite side to the mainsail, signalling the risk of an accidental gybe. Port tack is 'safer' in this respect than starboard as the wind generally veers as it gusts, which would bring it more on to the windward quarter.

Tacking and Turning in Strong Winds: It is easy to 'miss stays'—when the dinghy turns into wind, stops and falls away in the same tack. As soon as she stops, and makes it apparent that she is not coming round, reverse the rudder and put in a 'sternboard', that is steer the stern backwards to the position you were unable to gain by going forwards. But a known 'hard headed' boat can be induced to tack by the helmsman hardening in the mainsail the instant before going about and the crew 'backing' the foresail (holding it out to what is intended to be the new weather side) as the boat comes into wind.

Going from a run to close-hauled often results in missing stays as the helmsman over-estimates the amount of way his boat carries against a strong wind. It is better to take the manoeuvre in two stages—to go first on to a reach and trim the sails to develop maximum speed, then to come up into wind and complete the turn. This handling is also part of the manoeuvre 'tacking to avoid gybing', for having tacked, you then just bear away on to the new broad reach, with the boom now on the other side.

Gybing in Strong Winds: All winds have their lulls as well as their gusts and it is tempting to gybe during a lull. This may be just the wrong moment, as the gybe will be slow, you haven't a lot of steerage way and the next gust may hit you as you are halfway around. Better to gybe during, or at the end of a gust, when the boat is going fast (diminishing the effective wind strength if this is a run) so that the gybe can be effected smartly.

A light dinghy can be spun around with no adjustment to the mainsheet, if this involves also a change of course. At the command 'gybe-oh' the helm is put up and, as she spins, the crew helps the boom over while transferring his weight to the 'new' windward side, while the mainsail spills most of its wind in the process. A heavier, or longer dinghy will need to have her mainsheet hauled in before the gybe, to be let go as the boom swings over to cushion the slam and to prevent heavy contact between the boom and the leeward shroud. A similar technique is employed when a gybe is executed without change of course. The sail is sheeted in, the boom is handed across by the crew who, at the same time, transfers his weight to the windward side and while the helmsman pays out the sheet and applies strong rudder action to counteract the luffing effect of the sail on the new side.

The centreboard position for gybing in most dinghies should be about halfway; down enough to give control and up enough to allow some 'skid' during the manoeuvre.

Planing: Essential requirements for this are; a light

dinghy (for her size) with flat bottom sections aft; adequate sail area, correctly trimmed; a crew able to do this trimming while balancing the boat and a moderate-to-strong wind.

The fastest point of sailing, the beam reach, is also the best for planing, although a boat will plane on a broad reach or even a run, given enough wind. In marginal planing conditions, when the boat has to be coaxed to plane, it is necessary to wait for the gusts. As the gust is seen to be approaching from windward tactics may vary slightly according to which tack you are on. On starboard tack you can sit tight but on port tack it will pay to bear away slightly, as the probably veering gust may back-wind instead of filling the sails. As the boat accelerates the sails are sheeted in and crew weight is moved aft to help the boat adopt the slightly bow-up planing attitude. What is happening is that the dinghy is climbing on to her own bow-wave and leaving behind the stern-wave which is the speed-limiting factor of displacement boats. (See chapter 2.)

This planing attitude should not be exaggerated. If the crew moves too much weight aft the stern will drag and speed will drop. But if the fore-and-aft trim is too much down by the bow (and dinghies will plane on their forward sections for a little while) she is in an unstable attitude and the boat will sheer violently to one side or the other and come to grief.

Stronger winds raise the important requirement to keep the boat vertical, laterally. This is partly accomplished by the crew sitting out and balancing, and partly by the steering skill of the helmsman. As the boat begins to be overpressed and starts to heel, he steers to leeward slightly, 'runs away from the wind' as it were. Then, as the pressure subsides and the mast tends to lean to windward, he luffs to his former course. This is accompanied by paying out a little sheet on the first control and pulling it in on the second.

Light Weather: Maximum flow can be given to the sails by easing the luff and foot tensions, by moving the foresail fairleads slightly forward (or alternatively, raising the tack by fitting another shackle or strop), and by fitting more flexible battens in the mainsail. Softer control can be given to the sails by fitting thinner and more flexible sheets and by reducing the number of parts to the mainsheet purchase.

This is the time, too when 'wetted area', the parts of the hull, centreboard and rudder actually in the water, has a noticeable effect on boat speed and the smoothness of those parts is put in question.

Light-weather sailing technique is a combination of keeping the boat moving and keeping the crew still! In a race, which is when the problem arises, (otherwise you would paddle or row), it is worth putting in extra distance to keep her moving, while a few incautious movements, by stalling the centreboard or shaking the wind out of the sails, soon emphasise the importance of keeping the crew still. As when sailing into waves, so now too you get more drive if you sail a little freer, while keeping the boat upright makes the most of the wind supply and reduces leeway. On the offwind courses the rudder, as well as the centreboard, can be raised slightly to reduce wetted area.

The 'Roll-Tack': Manoeuvres and crew movement should be made smoothly during most dinghy sailing, and particularly in light weather, but the roll-tack is one instance when a little more movement can be allowed. Just before tacking the helmsman and crew use their weight to induce a heel to windward, this aids the turn, due to the asymetric immersed shape of the hull. Stepping across the boat as the dinghy nears the end of her tack, the crew 'sit her out' on the new windward side bringing her upright and sheeting in the sails sharply. This action, performed well, gives the boat an impetus for the new tack.

Heaving-To: The dinghy sailor soon learns that, to handle his boat well and competently, he must be able to sail her slowly (and even backwards at times) in addition to being able to obtain her maximum forward speed. Slow sailing he achieves by spilling wind from the sails, either by freeing the sheets or luffing into the wind, while downwind he can lower sails, trail his bucket —or lower the centreboard in shallow water!

With the wind forward of the beam, and usually on open water, there is the trick of heaving-to. For this the foresail is backed and cleated on the opposite side to the mainsail. The centreboard is kept fully down and the rudder is put down (tiller pointing to leeward) and the mainsheet pulled in sufficiently to 'keep her quiet'. This may result in a manoeuvre from the boat, with no further control, of alternately bearing away (as the balance shifts to the foresail) and luffing (as the balance returns to the mainsail). Slight mainsheet adjustment may result in perfect balance and the dinghy will 'forereach' slowly to leeward so, that, with adequate sea-room, the crew can get their lunch out! The technique can also be used, with experience, to bring a boat slowly into a leeshore berth, by forereaching up to it and raising the centreboard at the appropriate moment to increase leeway.

ANCHORING AND MOORING

If you sail inland you may see no need for an anchor

(unless it is the single fluked 'rond' or bank anchor) although it can still be useful to picnic or fish, from the centre of a lake or reservoir. In coastal waters however, this is essential equipment, for both safety and convenience.

Anchors

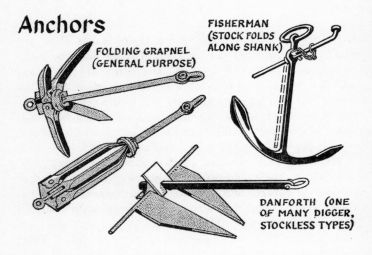

FOLDING GRAPNEL (GENERAL PURPOSE)

FISHERMAN (STOCK FOLDS ALONG SHANK)

DANFORTH (ONE OF MANY DIGGER, STOCKLESS TYPES)

In addition to the anchor you need a length of short linked chain (a fathom will do) a length of nylon or Terylene rope for a cable (strictly five times the length of the expected anchoring depth, but 10 fathoms is average) and a reel to wind it on and keep it tidy.

There are various types of anchor, which perform better, or worse, according to the nature of the sea bed— the digger types holding better in sand or mud—but the dinghy folding-grapnel type holds fairly well on most bottoms and has the virtue of being easier to stow. Weight is important, so match this with your size of dinghy, while the chain helps it to lie flat and dig in.

Dinghy LOA (ft)	Anchor (lb)	(kg)	Anchor cable—nylon or polyester, plaited	Breaking strain	Chain
8-10	3-4	2	¾ in circ	10 cwt	$\frac{3}{16}$ in
10-14	5-8	3-4	6 mm dia	500 kg	
14-16	9-12	4-5	1 in circ	18 cwt	$\frac{3}{16}$ in
			8 mm dia	900 kg	or ¼ in

Attach the cable to the chain with a 'fisherman's bend' and the chain to the anchor with a shackle. In use, the reel may be mounted in a bracket but do not expect it to take the weight of mooring. For this the cable must be cleated to a strongly mounted foredeck cleat or secured to the mast.

Mooring to the bank or quayside needs the minimum equipment of one painter. Do not be niggardly with this

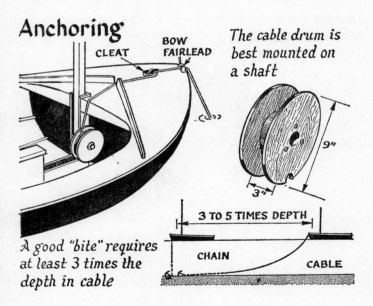

Anchoring

CLEAT

BOW FAIRLEAD

The cable drum is best mounted on a shaft

9″

3″

3 TO 5 TIMES DEPTH

CHAIN

CABLE

A good "bite" requires at least 3 times the depth in cable

Permanent moorings

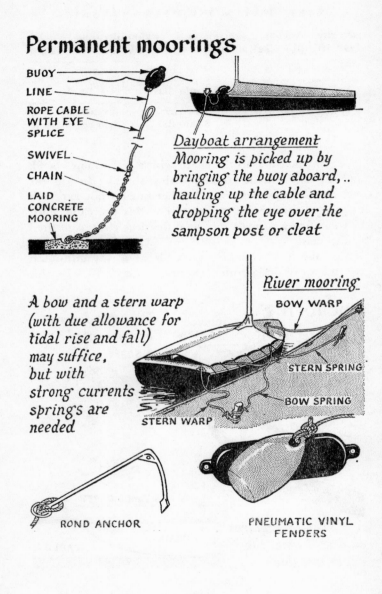

BUOY

LINE

ROPE CABLE WITH EYE SPLICE

SWIVEL

CHAIN

LAID CONCRETE MOORING

Dayboat arrangement
Mooring is picked up by bringing the buoy aboard, ..
hauling up the cable and dropping the eye over the sampson post or cleat

River mooring

BOW WARP

STERN SPRING

BOW SPRING

STERN WARP

A bow and a stern warp (with due allowance for tidal rise and fall) may suffice, but with strong currents springs are needed

ROND ANCHOR

PNEUMATIC VINYL FENDERS

and use one twice the length of the boat (you will be glad of this when the post is not as near as you would like it) and at least an inch in circumference—strong enough to be used as a tow-rope when required. The usual tie is a 'round turn and two half-hitches'.

If the boat is to be moored to the bank for a length of time she also needs a stern warp. This can be the mainsheet, but a rope kept for the purpose is better. If the boat is moored in tidal or swift-flowing water she will also need 'springs' (mooring warps which secure the bow to the stern mooring ring, and the stern to the bow mooring ring) with sufficient slack on all of them to allow for tidal rise and fall.

A fendered, or smooth, quayside is essential for the protection of your dinghy's paintwork if you are lying alongside. Failing that, or in addition, it is worth while dangling your own fenders. Two is the minimum number and the plastic, inflated ones are light, efficient and easy to stow.

9

Racing Your Boat

Racing is a Game: It is a pity that such a self-evident sub-title needs to be used, but to hear and read some people and some sections of the yachting press you would think it is the be-all and end-all of sailing. You either play it or you don't, but unlike other games like golf or bowls, where it makes no sense to use your equipment apart from the rules of the game, in sailing you can get just as much fun without reference to the racing rules— although there are rules of conduct and legal liabilities you cannot ignore.

But if you race you must abide by the rules and, in spite of the opening paragraph, I do recommend that you race. Not just once to 'see if you like it' (for most people have some off-putting experiences to start with), but occasionally throughout a season and interspersed with cruising, day sailing and class rallies (which are getting more popular these days and which also encourage the social side of sailing). Then, if the 'racing bug' bites, at least it will have done so in fair competition with the other 'bugs'.

One last thing. Try not to be put off by transitory appearances of unfriendliness and sharpness amongst racing people. They may sometimes look truculent before and during racing (competitive stress, no doubt) but they are mostly friendly people (especially after the race) and will welcome and help the newcomer all they can.

Handicap and Class Racing: Racing is conducted either under handicap, in which boats of all classes can race against each other, or as class racing, in which all boats have the same theoretical speed—they start together and the first one home is the winner. With handicap racing the theoretically fastest boat may be the 'scratch' boat and her 'elapsed time' over the course is recorded. Her competitors are given 'time allowances' (according to their theoretical speeds) and these times deducted from their elapsed times give 'corrected times' to decide the overall winner.

In practice, the scratch boat may also have a time allowance, but less than any of the others. These allowances are worked out from average recorded performances over many races and known as the Portsmouth Numbering System, or Portsmouth Yardstick and published by the RYA (Royal Yachting Association) who are the British governing body for racing. So if your dinghy has a yardstick of 97, for example, and completes the racing course in 97 minutes, another class boat with a number of 102 should take 102 minutes to do the same course in the same race. If that boat takes longer, then you have won; if she takes shorter, then she wins. Of course races take odd times to complete, not the convenient times quoted, and it is the race officer's job (equipped with a set of tables) to work out the fractions.

Equipped for Racing: The first thing is to have a boat which is acceptable to the club, regatta, or open meeting you wish to compete in; and the next, to see that she is properly equipped. Of course she must have a class certificate, as mentioned in Chapter 3, and if this is not in your possession you will have to get her measured by an official measurer and a certificate issued. The Class Association or the RYA will give you information as to procedure. If this is a new boat under construction the builder will make the arrangements. Sails are part of the measurement requirements and may have to be measured separately.

Ancillary equipment which falls under the inspection of the measurer, or race officer, includes the jib-stick or spinnaker-pole (checked for length), the self-bailers (checked for size and quantity) and the numerous 'go-fast' fittings which may, or may not, be allowed under class rules. But assuming all is in order, you have only to replace your cruising, triangular pennant for the square racing flag and you are ready.

RACE PROCEDURE

Competitors should arrive in good time to report to the race officer, to enter the boat's particulars and to read the race instructions. These show the course (with the number of laps to be sailed, which needs memorising or noting), the time of the start, and the code flags being used (each of which corresponds to a class or handicap race). Also indicated will be whether it is to be a gate or a line start.

Races are started by a sequence of simultaneous flag and sound signals (gun, klaxon or bell). The first comes as a ten-minute warning when the class flag is broken out with a sound signal. By then, you should be on the water in your boat with the sails set; you *must* be there before the next warning at five minutes, when the 'P' flag is broken alongside the class flag with another sound signal. By then you should be behind the line and should stay there until the start when the third sound signal is made together with the lowering of both flags.

Any boat over the line at the start (which means any part of her) is signalled according to race instructions, but usually by a second sound signal. It is then this boat's obligation to return to start correctly or be disqualified. Another signal, the First Substitute flag or further sound signals, means that too many boats were over for individual identification and there is a 'general recall', when the preparatory sequence of signals starts again. Race instructions may say that in this event a 'five minute rule' applies, which means that any boat sailing in the triangular area between the starting line and the first mark during the five minutes prior to the start pays the penalty of disqualification. This has the effect of keeping the boats well back from the line and ensuring an orderly start next time.

The Gate Start: The above describes the more usual line start but, in open meetings particularly, when classes can muster upwards of two hundred boats which could require a line difficult to discipline, perhaps half a mile long, the gate start is preferred.

This must be set in open water for the committee boat, which forms one end of the starting line, must be moored so that the first mark is exactly to windward. The 'gate' through which the entire fleet sails to start is opened by two boats; the 'pathfinder dinghy' which is one of the dinghy fleet but one who is (by known performance) capable of sailing a good close-winded course, and the gate launch.

One minute before the start the pathfinder dinghy starts sailing from the committee boat, close-hauled on port tack, towards the first mark. The gate launch, bearing the International Code flag 'G', takes up station behind her (keeping the same speed and distance). Three seconds before the start the launch releases a free floating

Racing starts

Line start

COMMITTEE BOAT
FIRST MARK
L/M
WIND
LIMIT MARK

Seconds to go and 21 will be "over" and must return outside the limit marks

Gate start

"G"

S/M

COMMITTEE BOAT
LAUNCH
FIRST MARK
WIND
PATHFINDER
START MARK (FREE FLOATING)
GUARD

GATE LAUNCH

PATHFINDER DINGHY (ON PORT TACK)

"G" FLAG YELLOW/GREEN

"U" FLAG RED/WHITE

GUARD BOAT

Competitors must start behind the gate launch, on starboard tack

buoy and, almost at once, the gate commences to open. Competitors start by crossing the stern of the gate launch and have the choice of starting early, when furthest from the first mark, or later, when they are nearer to it.

The gate can be extended as long as it is necessary to let through all the fleet; the last to go being the pathfinder dinghy herself, who gybes round and also passes behind the gate launch and joins the race. There can be another boat involved, the guard launch, which 'sweeps' the fleet clear for the pathfinder dinghy.

WHEN BOATS MEET

It should be noted that the pathfinder dinghy in the gate start is on port tack, and yet she has right of way over every other boat in the fleet. This is one of the few times when a boat on port tack has priority and this is done so that boats starting behind the gate launch do so on starboard tack, the normal right-of-way tack, and their start can be unchallenged.

The rules of racing are extensively propounded in publication YR1, obtainable from the RYA, and no one who races should be without a copy. What is written here is not a substitute for the rule book, but a few simplifications and generalisations which should give you the idea and keep you out of trouble. Fundamental rules may be mentioned, but there are situations when even these are not inviolate.

The rules apply from before the start, when the boat intending to race sails in the vicinity of the starting line, until after the finish and clear of the course.

The Opposite Tack Rule: *A port tack yacht shall keep clear of a starboard tack yacht.* This rule applies whether yachts are racing or not, as it is also a rule of the Inter-

Rights of way

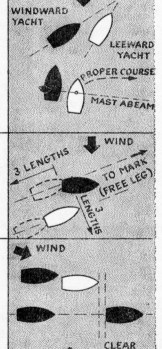

Opposite tacks ~ port gives way to starboard

Same tack ~ windward gives way to leeward. (white may luff black; head to wind, if she chooses)

Curtailment of luff ~ only when windward helmsman can call "mast abeam"

Sailing "below" a course, when less than 3 lengths ahead or to windward of another yacht passing to leeward, is illegal

Overtaking yacht must keep clear.

Overtaking to leeward ~ the yacht must give the windward yacht room and not sail above her proper course (luff) until clear ahead

WIND
PORT TACK
STARBOARD TACK

WIND
WINDWARD YACHT
LEEWARD YACHT
PROPER COURSE
MAST ABEAM

3 LENGTHS
WIND
TO MARK (FREE LEG)
3 LENGTHS

WIND
CLEAR AHEAD

national **Regulations for Preventing Collisions at Sea.**
It applies too, whether beating, reaching or running. The
port tack yacht must bear away and go behind the yacht
on starboard tack, or tack before they meet. This must
be done in such a manner that the right-of-way yacht
needs to take no avoiding action, or the offending yacht
can be disqualified. The right-of-way yacht has her
responsibilities too; not to alter course so as to prevent
the other yacht from keeping clear.

The Same Tack Rules: *The windward yacht shall keep
clear of a leeward yacht*: This rule is in the International
Regulations too and means that you are not allowed to
bear down on another dinghy which is converging on
the same tack and that if a dinghy to leeward deliberately
points higher you must respond and keep clear. This
action is known as 'luffing' and is employed to stop an
overtaking boat from passing. But there are curtailments
to this luffing ploy. A yacht may not luff another above
her proper course if, when sighting abeam from his
normal position and sailing no higher than the leeward
yacht, the helmsman of the windward yacht is abreast
or forward of the mainmast of the other. And there are
restrictions about luffing before the start, when a luff
must be carried out more slowly.

The windward helmsman may curtail the luff by calling
'mast abeam' when the luffing yacht no longer has luffing
rights and that yacht must immediately resume her
proper course.

*A yacht clear astern shall keep clear of a yacht clear
ahead*: This refers to overtaking and is mostly applicable
on the offwind courses and especially on a run, when the
overtaking yacht is 'blanketing' the wind from the yacht
in front with her own sails. It must also be remembered
at marks, where boats naturally converge and where

special rules have been formulated to deal with tight situations.

Changing Tack Rule: *A yacht which is either tacking or gybing shall keep clear of a yacht on a tack*: This means that a yacht, tacking from port on to starboard tack and so giving herself 'right of way', is prevented from doing so in front of a yacht which is following on port tack and slightly to windward, unless this is done, and the tack completed, before the following yacht must take avoiding action. It also applies when a port-tack yacht tacks closely in front of a converging starboard tack yacht.

ROUNDING AND GIVING 'ROOM' AT MARKS

A mark must be rounded in the direction stated in the sailing directions (either leaving it to port or to starboard) and no part of a boat or her crew must touch it. The penalty for doing so is to round it again, which means making a complete circuit, and this time without touching, before continuing.

When two, or more boats round together the 'outside' boats must give the 'inside' boats room to round if they are entitled to be there. No entitlement is possible if they are on opposite tacks when beating, for the rule concerning this is the same as if the mark were not there. No entitlement is granted if the inside boat has not established an 'overlap' on the boat, or boats 'outside' her. An 'overlap' occurs when the bow of the training boat is forward of an imaginary line projected abeam from the aftermost point of the boat leading her, and to claim 'water' at the mark, this overlap must have been established before two boat's lengths from the mark. The entitlement having been properly gained, still applies if

Keeping clear

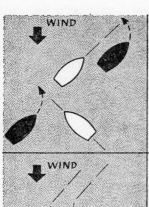

Tacking must be done far enough away from another yacht "on a tack" and not obstruct her, before the tack is completed

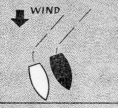

Gybing ~ when two yachts are doing so together the one on the other's port side must keep clear.

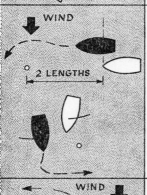

"Room" at the mark ~ must be given to an inside overlapping yacht if the overlap occurred *before* two boat lengths from the mark

This applies also to opposite tack yachts at a downwind mark

No room is allowed to a port tack yacht at the weather mark. (Opposite tacks rule applies)

the overlap is subsequently broken, that is, within this two lengths.

'Room' to clear Obstructions: Safe pilotage is an essential ingredient of the racing rules, indeed, it could be said that they are designed not so that one boat can take advantage of another but so that numbers of boats can navigate together in a confined area without hitting one another. In any situation there should be the onus on one boat to hold her course and on the other boat to take avoiding action. However, I am not going to say that advantage is never taken but, by and large, the rules work very well.

This safety aspect is catered for when there is danger of a yacht being 'run ashore' or fouling an obstruction. When two close-hauled yachts are on the same tack and the one to leeward or clear ahead must tack or make a substantial course alteration to clear a danger, but cannot do so for fear of colliding with the other yacht, she may hail—'room to tack'. The windward or following yacht will then either call 'you tack', and steer behind the hailing yacht or herself tack, thereby giving the yacht in danger clear water.

Infringements: When a rule is infringed, and this is owned by the culpable boat she should retire. However, the sailing instructions may specify that a boat may exonerate herself by a penalty. For dinghies this is the 720 degree turns. The dinghy must do two complete turns in the same direction but keep out of the way of other boats while doing so. This may mean going slightly outside the course. If the infringement involves another boat, this boat must be informed that the penalty turns are being performed. After the race the details of the incident and the penalty must be given to the race committee.

Penalties

Re-rounding the mark correctly exonerates a yacht hitting it the first time. The faulty first rounding must be completed first

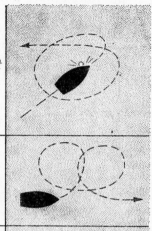

Two 360° turns may (under certain conditions) exonerate a dinghy that has infringed a rule

Obstruction

When a yacht, sailing into an obstruction, say, shoreline, cannot tack without colliding with another boat she may hail "water", when the other boat, or boats, must tack to give room

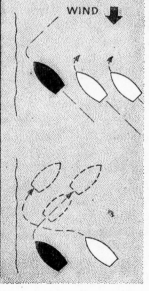

WIND

Or one hailed yacht may reply "you tack", when the onus is then on her to keep clear. She may prefer this to get a favourable shore "slant"

If the rules always worked perfectly there would be no disagreements and no need for protests. But sometimes not only do boats collide while both parties believe they are in the right, but the rules appear to collide too! So the boat who believes she has been offended against, which sometimes means both of them, flies the protest flag, which is a white square one and attached to the shroud. At the same time notice is given to the offending boat that protest action is being taken. The other boat may then decide to acknowledge responsibility and retire, perform the penalty turns or decide to continue and contest the protest. If she continues she must be given her usual rights of way by the offended boat and all others.

The sequel is the notification to the race officer, which must be done as soon after the race as possible, and the protest hearing. At this a committee is presented with the case by the protesting boat's crew, the evidence of witnesses and the defence. Sooner or later, they come to a verdict and the protest is upheld or dismissed with appropriate disqualification.

BEGINNING TO RACE

On the 'don't run before you can walk' principle no one would expect to go out and sail competitively right away. On the other hand it would be wrong, and only throw the fleet into confusion, if the newcomer 'gave way' whenever she met another boat. A race can be sailed cautiously, however, by steering clear of 'mêlees' and by knowing and abiding by the rule interpretations which have just been given. This means sailing with the 'back-markers'; but gradually experience will grow, and sailing will become more competitive.

Start well behind the line and watch the starting techniques of the others then, if the first leg is a beat, be sure to approach the first mark on starboard tack. This is a good policy for old and new hands alike for it gives 'right of way' where boats must converge. Approaching on port tack will see you being 'put about' by starboard boats and in all sorts of trouble.

The next leg will probably be a reach and it may be possible to keep well up to windward. This means that if there is any crush around the 'wing' mark you need not get involved but can sail right around the outside of mark and boats. Finally, there may be a broad reach or a run, and again, you can steer towards the outside of the course to enable you to give the leeward mark a wide berth if you want to.

Around the course you can be learning by watching the tactics of others, but not so closely that boats, perhaps converging, in the immediate vicinity are overlooked. On the beat, for example, keep a look-out for approaching port-tack boats and call 'starboard', to warn them of your presence. Then maintain your course so that each boat can take avoiding action. If you are on port tack, however, also keep a watch, rather than wait for a call, for this warning is not obligatory. And on a reach, if a boat appears to be bearing down on you while you are sailing the proper course, be sure to call 'loo'ard boat' and assert your rights.

Sailing to Win: There are many books and yachting magazine articles written on this subject and the serious reader must be referred to them rather than a few paragraphs that are all that can be crammed into this introductory sailing book. Pundits are numerous and there are almost as many formulae for success, often based on improving one's boat, equipment and sails—which must be good for the commercial side of yachting. But most would agree that the helmsman and crew are themselves the most vital ingredient in any formula, with their ability to get the best out of their boat and, at the same time to be able to look around, sail the course intelligently, read the wind shifts, cover the other boats and—well, here we go again!

Unless you are 'top notch' something has to go and I would recommend that this is never the sailing of the boat. A calm, unflustered approach to this is essential and may not be possible if you are worrying too much what the other boats are doing or whether your equipment is right. And this is how to enjoy your racing too.

The Line Start: A watch is essential to make a good start and there are specialised yachtsman's stop watches, with a sector indicating minutes to go and a long sweep seconds hand, which are ideal. Set the watch at the ten minute signal and check it again at five. If there is time, examine the line and sail the first leg before the start. If this is a true windward leg there may still be a 'bias' in the line, that is, not at right angles to an imaginary line from the centre of the starting line to the first mark.

This is often 'arranged' by race officers. Since starboard is the right-of-way tack it is favourite to start at the starboard end of the line, and most boats do, resulting in a huddle and infringements. By slanting the port end nearer the first mark the boats are more inclined to spread themselves along the length, while there will be the bold boat or two willing to chance a port tack start at the port end of the line, with the hope of 'clearing' the whole fleet, then tacking on to starboard to 'cover' them.

Some lines are so biased or the wind direction is so far from providing a true windward leg that it is more important to start at the right end of the line than to get there on time! Getting the boat moving fast is of equal, if not greater, importance than timing and if there is a huddle of boats at, say, the starboard end, each trying to get the windward position, they are slowing one another down. A boat sailing in clear air and water behind the line may well be able to sail fast at the line and cross it near the mark, five or ten seconds after the 'gun' and 'shoot' into a very good position.

When the mark, indicating the end of the line, is surrounded by navigable water, it is possible to get 'squeezed out' by another boat or boats. There is a rule covering this which is an extension of the ones concerning giving room at marks of the course and the right to luff.

Before the start all boats are governed by the normal rules or rights of way and luffing (except that they are required to act more slowly). But after the starting signal a boat may not deny a windward boat room at the starting mark either by luffing higher than close-hauled, if this is a beating start, or higher than the direct course to the first mark, if this is a reaching or running start.

A Clear Wind: When sailing dinghies are so equally matched it is especially important to get clear wind for one's own boat. This is why there is always bunching at the start with boats competing for the windward position, but it is equally important on the beat, and on others legs as well.

A yacht not only derives her motive power by interrupting the air stream (wind) with her sails, she also deflects and causes turbulence in that air stream. This is collectively called 'dirty wind'. It consists, when close-hauled, of deflected air *to windward* (from a line about 30 degrees from the centreline, aft from her bow) and changing gradually into turbulence (to a line about 45

Starting and tactics

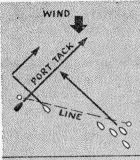

A starting _line "bias"_ prevents "bunching" at the favoured starboard end. A "classic" port tack start may be baulked by a star-board tack boat at that end

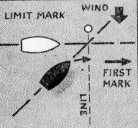

A starting _line mark_ limits the line length. In this example White must keep clear of Black before the start gun, but she can claim "room" at the mark, after

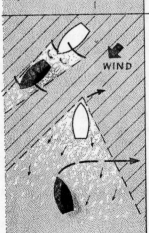

"Dirty wind," the disturbed air from a boat's sails, is best avoided ~ if possible. Downwind, this is called being "blanketed" and is difficult to escape from. Close-hauled, the remedy is to tack but the boat ahead may also tack. This is known as "covering"

degrees on the other side of the centreline, and extending for two or three 'sail heights' to leeward). The result of this on neighbouring boats to windward or behind is to 'head' the wind for them, causing them to fall off to leeward, and, for boats to leeward, to entail lost drive, causing them to fall away even further.

So if you are 'caught' in either of these vulnerable areas it would be advisable to tack to 'clear your wind' at the first opportunity. But if this is not possible, as when there are a lot of starboard-tack boats to windward, it is better not to fret. Instead concentrate on sailing your best with the wind you have; the result will be less disastrous.

Safe Leeward Position: The windward position is a very strong one to be in but the 'safe leeward position' is as good, if not better. Here, slightly forward and to leeward, you are able to deal out deflected wind and the boat to windward will find it impossible to pass or gain. Particular occasions when it can be employed are at the starting line and when, on the beat, you are 'put about' by a starboard tack boat and are able to tack just in front and to leeward but not so close as to require a change of course from her before you have completed the tack!

Clear Wind on a Reach: The anxiety of boats to have a clear wind on this course, (for it is very easy to pass to windward of another dinghy and almost impossible to pass an equally matched one to leeward) tends to have them sailing to windward of the direct course to the next mark. 'Luffing matches' occur frequently which drive them further to windward. These boats may then find themselves having to free-off to broad reach down to the mark—a slower point of sailing.

Clear wind in these circumstances may be not to windward but to leeward of the direct line. A boat taking this 'low' course will have wind which has had time and distance to pick up after being troubled by the boats to windward. Furthermore, when the boats to windward are broad-reaching slowly down to the mark, this leeward boat is on a faster point of sailing, close-reaching up to it.

Calms and Wind Shifts: When the first leg of the race runs parallel to the shore, as it is bound to do on a river and may do on a lake or on the sea, the question arises as to whether more wind is to be had inshore or offshore. If the wind is blowing offshore the boats at the start line jostling for windward berth may be luffing each other into an area of calm. In this case it often pays to accept the leeward position, with the bonus of more wind.

During the course of a race, care should be taken not to sail into one of these calms. Invariably there will be such an area near the windward shore in light winds, but if it is very light it may well be further off-shore, say, in the middle of a lake. The appearance of the water is a good guide, which can be confirmed by the performance of other boats. In this very light weather the thermal winds near the shore may be the strongest winds to be had.

As I said earlier, wind is always varying in strength and direction. This has its greatest effect on dinghies sailing a race on the windward leg, and the helmsman who can profit most from these vagaries will pick up many places. Given a free choice (the presence of other boats could be a limiting factor) one starts off this leg with a long board on the more profitable tack but without getting too far from the direct line between the two marks. One then tacks and does a shorter board to the other side, then a shorter one and so on. The purpose of these shortening-tacks approach is to be able to take

Racing techniques

The safe leeward position
dinghy ~ the "slightly ahead
and to leeward" one of two
close-hauled boats~delivers
backwinded air to her rival

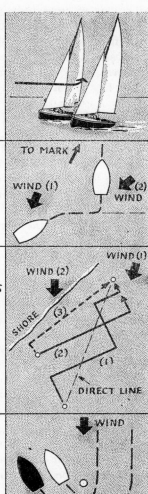

Tacking in a windshift,
when beating, shortens
the distance sailed to the
weather mark

Choice of tacks. If it is a
"plug" dead upwind (1) tacks
of diminishing length is good
policy. If it is a "dog leg" (2)
put in the longer tack first,
unless there is a shoreline
when it may pay to go
looking for a freer "slant" (3)

Rounding the leeward mark,
by leaving it wide on the
approach and cutting it
close on the turn, often
earns a windward position

advantage of any wind shifts on the way.

A shift of wind is best detected by shore transits if you are within sight of the shore; offshore, sea marks or a compass can be used. Other pointers are the performances of other boats, smoke and the tell-tales of one's own racing flag or cottons tied to the shrouds.

It is not advisable to tack on every shift, only the major ones, but detecting those is not easy. They can be, to some extent, anticipated in the gust-and-lull sequence noted in Chapter 8, and there will be a fair chance of one near a windward shore. Here it often pays to continue into the 'headed' area of wind before tacking and taking full advantage of its slant.

Current and Stream Considerations: When the windward leg runs parallel to the shore there may be a tidal stream which may make the offshore and leeward situation described earlier not so clear cut. If this is the case it almost always pays to keep in the slacker water close to the shore or the bank. This may be imperative in the case of a tidal estuary when, to round a windward and up-tide mark in the fairway, it is necessary to tack way past it along the bank, and then reach out to it.

The Downwind Leg: As the dinghy is now going the same way as the wind this reduces the apparent wind speed and makes this the quietest point of sailing. Yet many races are won and lost on this leg. If you don't have a spinnaker your sails should be set goosewinged, if possible, and the use of a jib stick assists this. The foresail may even be set to windward when the wind is on the quarter, making it into a small reaching spinnaker.

Both kicking strap and the mainsail clew, if this is adjustable from within the boat, should be eased to give more drive and the centreboard raised to reduce wetted area. Helmsman and crew can sit either side on the side-decks, not only to give better balance, but because their 'windage' is as helpful now as it was harmful when on the beat.

Tactics should include trying to avoid the 'blanketing' effect of the boats coming up astern while being prepared to 'play' the turbulence of one's own sails on the boats ahead. The effect of this is devastating and often results in 'luffing matches'. In my experience these give little benefit to either boat but usually assists the third boat coming up astern. Yet it may be a necessary tactic if this is the last leg of the course.

Normally, it is better to let the overtaking boat pass without demur and then, by partially blanketing, keep her within 'striking distance' until fairly close to the leeward mark. There may then be an opportunity to establish an overlap and claim 'water' at the mark. Or she may try to defend herself against this, and sail so close to the mark that she swings wide on the windward leg, while the following boat can take the mark wider and come up to windward of her.

SPINNAKER, THE DOWNWIND SAIL

I made little reference to this sail in Chapter 4 as this is primarily—though not exclusively—a racing sail.

The sail is usually as large as the combined areas of the working sails but the fact that it is used downwind, lessening the apparent strength of any wind, makes this less formidable. Nylon is the ideal material, for its light-ness and 'stretchability' allow it to be 'blown into shape' by the wind rather than holding a fixed shape of its own. Nevertheless, you get different 'cuts', from the very full 'balloon' type suitable only for running to the flatter 'reaching spinnaker' for use as its name suggests. Running could include an apparent-wind direction of within a sector of 15 degrees on either side of the dead aft centre-line but, depending on the cut, the sail may continue to work with the wind further forward, while the reaching spinnaker will pull with the wind forward of the beam. Performances will also depend on wind strength.

The working parts include a halyard, which can be either internal or external to the mast, with a swivel snap shackle for quick attachment to the head of the sail, while the hauling end is led to a cleat inside the boat—preferably aft within reach of the helmsman—who can yank on it while the crew is feeding out the sail or managing the pole.

Guys, which can be of a fairly light rope, are permanently attached to the two clews. These run through fairleads, positioned by experiment but invariably a long way aft, and then to jamming cleats—although they will often be 'played' by hand. An uphaul/downhaul vang, (consisting of a composite length of which the lower is polyester cordage and the upper, for tensioning, is elastic shock cord) runs down forward of the mast from the hounds to the deck. Completing the essential equipment is the pole, or spinnaker boom. Its length is decreed by class rules, but it should be long enough when fitted and pointing forward, for the spinnaker clew to be clear of the forestay. The fittings are 'double ended' or duplicated, so that you don't have to search for the right end, and to facilitate gybing. These consist of a hook at each end, (one for the clew and one to engage in an eye on the mast) and a double-sided cleat at the mid-way point, for the vang. The hooks can be the cheaper open ones but piston hooks, actuated by a cord from one to the other, are better.

The Spinnaker Chute: This optional extra may be fitted

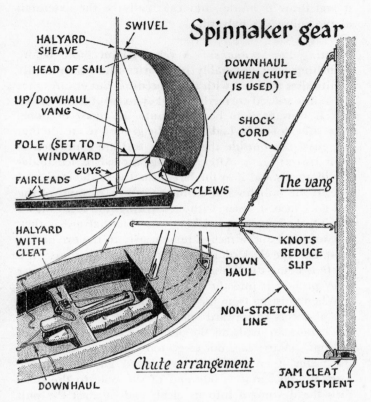

Spinnaker gear

SWIVEL

HALYARD
SHEAVE

HEAD OF SAIL

DOWNHAUL
(WHEN CHUTE
IS USED)

UP/DOWHAUL
VANG

SHOCK
CORD

POLE (SET TO
WINDWARD

GUYS

FAIRLEADS

CLEWS

The vang

HALYARD
WITH
CLEAT

DOWN
HAUL

KNOTS
REDUCE
SLIP

NON-STRETCH
LINE

Chute arrangement

DOWNHAUL

JAM CLEAT
ADJUSTMENT

inside the boat, with the mouth of its large diameter, rigid plastics tube emerging through the foredeck near the bow and extending aft through the buoyancy tank. Alternatively, it lies along the deck, but in both cases is continued aft by a fabric tube, held open by externally fitted hoops. The only additional fitting may not be one at all, for the spinnaker downhaul (not to be confused with the downhaul vang) is a rope attached to the centre of the sail and led down through the chute, (to collapse it and draw it inside), but can really be the extension of the spinnaker halyard.

Hoisting the Spinnaker: A rehearsal on shore before launching, and preferably in calm conditions, not only familiarises the crew with the procedure, but ensures that the sail is stowed correctly—the key to trouble-free hoisting. If stowage is to be in a chute this *must* be done. The sail can first be laid out on the ground to ensure that the guys pass outside the shrouds and are connected to their correct clews. After hoisting, the sail is led under the jib-sheets and into the boat, usually on the starboard side. If the sail stows in a bin the centre, or bunt of the sail goes in first, then, without twisting, the rest follows, leaving the two clews sticking out. The halyard is then disconnected and attached temporarily to the eye on the mast, and the slack on the ropes tidied up. Of course, if a chute is being used, it should be necessary only to house it by pulling it into the chute with the downhaul.

When hoisting from the bin the halyard is taken under the jib-sheets to leeward and connected to the head. As the helmsman hauls on the other end the crew feeds the sail out, (taking care not to go too fast, or it will fall in the water) then draws the windward guy around the forestay and connects one end of the pole to the clew, slips the downhaul into its cleat, and (against the pull of the shock-cord) pushes out the pole and clips the nearer end to the eye on the mast.

Trimming the Spinnaker: Although this is, to appearances, a 'bag of wind' there is invariably a flow of air across the sail from the luff (which is the side the pole is fitted) to the leech (the opposite one). The guy attached to the leech clew is called the sheet although, if the spinnaker is being set on the other side, one has to remember that luff, leech, guy and sheet all reverse their names.

This sail, like the fore-and-aft ones, has a point of maximum pull and the sail is best trimmed when its

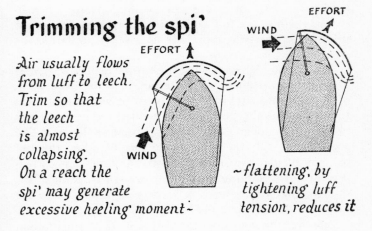

Trimming the spi'

Air usually flows from luff to leech. Trim so that the leech is almost collapsing.
EFFORT
WIND
On a reach the spi' may generate excessive heeling moment –

EFFORT
WIND
– flattening, by tightening luff tension, reduces it

pull is furthest forward. This is where using the sail on a beam reach can be self-defeating—the pull comes too much to the side and develops more sideways heeling force than forward drive.

For the initial setting both guy and sheet must be adjusted together, pulling the pole to windward and the foot around the forestay. Pole adjustment is controlled angularly with the guys and horizontally with the vang. Angularly, it needs to be as far out from the centreline as it will go without causing the luff to collapse; this gets the sail away from the blanketing effect of the mainsail. Horizontally, in which plane alterations are made manually at the vang cleat or by increasing or decreasing the shock-cord tension in a cleat at the deck, it straightens or curves the luff to suit the amount of wind. In very light airs, when the sail is hardly filling at all, a downward angle sometimes helps as movement 'shakes' what wind there is out of the sail. With a little more wind it can be raised, which lifts the spinnaker away from the turbulence of the mast and into the clearer air. This can be accompanied by an easing away of the halyard. Then in stronger winds, the luff can be straightened by downward pressure on the pole, especially on a reach, to flatten the sail.

One cannot be specific about spinnaker control; it is a sensitive instrument which responds to the inspiration of the crew, but a typical broad reaching adjustment gives the idea. Pull the pole around with the guy until it is roughly at 90 degrees to the wind, then make your fine adjustment with the sheet. Ease this to allow the luff to come further to windward, until the time comes when it is all but collapsing, then sheet in slightly. Remember, though, that trimming is a continuous task, due to the variations in the wind and boat's speed.

Whether the crew does the trimming or the helmsman does it depends on who sits to windward, the only place for a good view of the sail. On a broad reach or a run the helmsman often sits to leeward, enabling the crew to sit well out to windward—even on the trapeze. Alternatively the helmsman may prefer to sit windward, especially in light weather, where he can helm and 'tweak' the guy. On a reach, when both are sitting to windward, it is best left to the crew.

Gybing the Spinnaker: Gybe the mainsail first and settle the boat on her new course which is presumably also a broad reach on the other tack. The crew then disengages the pole from the mast and attaches it to the 'new' weather clew so that it is supported between the two clews. Next, he disconnects the pole from the 'old' weather clew and fits it to the mast and, finally, adjusts sheet and clew.

This, and the various acts of spinnaker management, are best done, and with less risk of twisting or 'sailing over it', if the guys are not given too much rein and the clews are kept, as much as possible, at right angles to the wind.

Downing the Spinnaker: The reverse of the hoisting procedure is for the crew to remove the pole and pass it back into the boat (trying to avoid the helmsman's eye); then that 'worthy' (who should be able to steer with his knees while his hands are busy) uncleats the halyard and pays it out as fast as the crew can smother and gather in the sail. This he feeds under the jib-sheets and, bunt first, stuffs it into the bin, leaving the clews poking out. There remains only to disconnect the halyard, clip it to the mast and secure the pole in its stowage.

FAIR AND FOUL MEANS OF PROPULSION

It should be self-evident that races should be sailed and boats propelled only by the action of the wind on the sails. Yet this needs to be written into the rules, apparently. A paddle may be used, but only for steering, in an emergency. 'Sculling' with the rudder, by working it from side to side would be frowned on.

Other proscribed methods to give impulsion are 'ooching' and 'rocking', the former being the action of the crew of moving their weight forward slowly and stopping abruptly, and the latter of rocking from side to side. 'Fanning', which is moving the sails in and out by manual power 'like a bird's wing' is also discouraged while 'roll tacking' (described in Chapter 8, page 208) is illegal if performed frequently in calm conditions.

Mooring is all right but not kedging, which would be throwing the anchor ahead and moving the boat by hauling on it. Mooring includes having a crew member going overboard to stand on the bottom and, if the boat has gone aground or needs manual power to clear an obstruction, he may push the boat off, either from the bottom or from the shore.

THE FINISH

The finishing line is described in the race instructions and care should be taken to finish on the correct side of any limit mark. The same rules apply to these as apply to marks of the course. Again, it often pays to make this last approach on starboard tack, if this is a windward last leg, to avoid being 'put about' on the line. A yacht has finished when any part of her hull crosses the finishing line; she may thereafter clear the line in either direction.

The Declaration Sheet: This should be signed by the competitor, either to declare that he has completed the course and sailed 'by the rules' or to notify his retirement. When race courses go out of sight of the race officers, this sheet is used as a check by them to ensure that all of their competitors have returned safely.

FLAG SIGNALS

Instructions to competitors are given by flag signal. Here is a selection of some of the more commonly used ones. An adhesive key to the International Flags should be bought and stuck in the boat.

'AP' Answering Pendant—Postponement signal. The

warning signal will be made one minute after this signal is lowered, or postponed the indicated number of hours, if flown over a numeral pendant.

'*N*'—Abandonment signal, meaning 'all races abandoned'.

'*N over First Substitute*'—Cancellation signal, meaning that all races are cancelled.

'*P*'—Preparatory signal (the Blue Peter), meaning that the race will start in five minutes exactly.

'*S*'—Shorten course signal, meaning (usually) that the race will finish at the end of the round yet to be completed by the lead boat.

'*First Substitute*'—General recall.

10

Handling Your Boat Ashore

CARRYING YOUR DINGHY

Inevitably the dinghy which looks after you best afloat, the bigger one that takes less agility and effort and can tolerate rougher conditions, is the most awkward one to handle on the shore. Moreover, she needs more looking after in the way of maintenance. These facts often have a bearing on the choice of dinghy, the choice of where to sail, or both.

Things to bear in mind, as regards handling are: the manual power you can summon; the mechanical aids you need and the firmness and accessibility of the approaches.

With plenty of fit, willing helpers, a lot of 'boat' can be humped on to a car roof, lifted over poor approaches and launched or recovered from rough water. But a man alone, perhaps with the aid of his wife or a child, must use mechanical devices and choose only the best approaches to the water or he must limit the size of his boat.

The Roof-rack: This means of transporting a boat has the attractions of avoiding the bother of a trailer, of keeping the equipage within the car's length, or little more, and of not being affected by a trailing vehicle's speed limits.

Car manufacturers are not so keen on the idea,

averring that the structure of the vehicle is not intended for such stresses, the centre of gravity is raised to the extent of causing harmful roll characteristics, and the windage, apart from increasing petrol consumption (which is the owner's worry), affects the steering.

Nevertheless people will 'roof' so some notes will not come amiss.

The weight ratio between the car and the boat is the most important factor and the greater this is in favour of the car the better. But for the medium-sized car a boat weighing 100 lb should be the upper limit. This takes care of surf-boats and dinghies up to the little Mirror. Big cars may take up to an Enterprise or a Fireball—but you still need the manual lifting power to get them up there.

The type of roof-rack to employ is that which attaches to the gutter of the car roof. The ones with pads may leave dents, and always leave 'rings'. The boat should be tied down to the car and not just to the roof-rack. This means ties at the bow and stern, where the overhangs occur. This also reduces 'pitching'.

Some manufacturers have designed ingenious roof racks, incorporating loading ramps, enabling one man to load the small dinghy on to the car roof. Some also do double duty as launching trollies—for one of the snags with 'roofing' is that you still have to get the boat to the water so that you may need to carry some sort of trolley.

The Road Trailer: This is the choice for most dinghy users. Advantages are the low centre of gravity for towing, a low loading height, the possible fitting of overrun brakes (a legal requirement when the unladen trailer weight exceeds 2 cwt but not usually fitted otherwise) and the fact that the trailer can be unhitched and hitched quickly at the launching ramp.

Car roof transport

GUTTER
MOUNTED
RACK

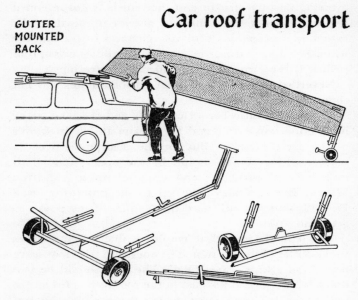

The "Christy, Sololift Launcher/Loader" ingeniously aids unloading then, with a quick assembly, converts to a trolley. Dismantling makes for easy transport

The disadvantages are centred around the manoeuvring and parking difficulties of the articulated length when driving. You are also called upon to pay double dues at many toll bridges, roads and tunnels; on ferries, too, which is to be expected. The ones who do not expect this double charge are those who carry their boats on the roof!

The capacity of a trailer is determined first by its length, and then by its springing. Specifications for these are given by manufacturers in their brochures. When arriving at one's own requirements allowance must be made for the extra weight of mast and spars, accessories and all the odd pieces of luggage which are apt to find their way into the boat, thus increasing the load. So it is better to err on the generous side when choosing trailer size, which will then tow more steadily. On the other hand, if yours is a light boat she will get a 'hard' ride if she is put on a trailer designed for a heavy load. Not only is a light trailer better for trailing a light boat but the trailer, with or without the boat, will be easier to manoeuvre manually.

Types of springing include the leaf (one manufacturer, Wicksteeds, make it possible to up-rate the capacity of their trailer by fitting extra leaves) the steel coil, rubber and the torsion bar. Leaf and coil springs need some maintenance but rubber and the torsion bar are maintenance-free. This fact is worth remembering if you intend to use your trailer as a launcher.

Rubber springing can be in the form of the Flexitor Suspension Unit, in which the axle arm is embedded in solid rubber enclosed in a steel box, or the 'rubber ball' type, which cushion between the trailer axle beam and an extension of the wheel axle, which has a limited hinged vertical movement. The Aeon Hollow Rubber Spring is a suspension unit of this type.

Trailer Design: There are two basic forms; the backbone, and the 'A' frame. Both have advantages and among those of the former are simplicity and that the boat can be supported along the whole of its length (if the draw-bar is long enough), adjustable chocks, or rollers, being placed just where they are needed. But there is another important feature in that the axle beam can be made adjustable along that length. This allows the correct amount of imbalance which, for safe towing, should weigh *down* on the car-coupling from 30 lb, for

Road trailers

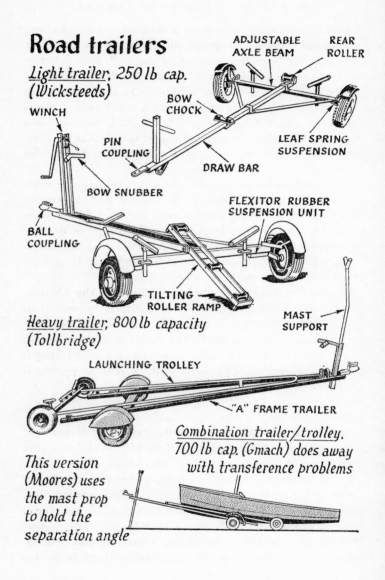

Light trailer, 250 lb cap.
(Wicksteeds)

ADJUSTABLE
AXLE BEAM

REAR
ROLLER

WINCH

BOW
CHOCK

PIN
COUPLING

LEAF SPRING
SUSPENSION

BOW SNUBBER

DRAW BAR

FLEXITOR RUBBER
SUSPENSION UNIT

BALL
COUPLING

TILTING
ROLLER RAMP

Heavy trailer, 800 lb capacity
(Tollbridge)

MAST
SUPPORT

LAUNCHING TROLLEY

"A" FRAME TRAILER

Combination trailer/trolley,
700 lb cap. (Gmach) does away
with transference problems

_This version
(Moores) uses
the mast prop
to hold the
separation angle_

a light equipage, to 50 lb, for a heavier one.

If this adjustment results in the axle-beam being further than one third forward from the rear end, the trailer is really too long, while if the axle-beam is at the extreme end of the draw-bar there is excessive overhang for the boat, and the trailer is too short. Some designs incorporate a telescopic draw-bar which accommodates this.

With the 'A' frame trailer the weight of the boat is taken at only two points—at the bow and at the lateral member at the rear, or the base of the 'A'. The advantage of this design is in its intrinsic strength and the fact that no 'whippiness', sometimes present in the backbone type, is possible. The boat can be very well chocked, not only at the axle beam but also along the rear lateral. The type lends itself to fore-and-aft leafspring suspension, which gives a comfortable 'ride' and the main disadvantages are heaviness and the fact that it is difficult to arrange an adjustable axle. Usually you are left to the expedients of distributing the boat along the length or disposing the movable weight inside the boat, to get the required imbalance.

Chocks and Rollers: Perhaps the best support for a boat on a trailer is similar to that designed for the Turtle dinghy by Thames Marine. This is a resinglass moulding 'tailor made' to cradle a complete lower half-section of the boat. I have also seen this made up in plywood for other boats. The difficulty arises when loading and unloading—the boat must be lifted 'in' and 'out'.

Most trailers have adjustable padded chocks, which support the boat at the bilges, while Snipe Trailers prefer large rollers. This is all right so long as the weight of the boat is taken down through the keel—if the chocks are made to be load-bearing there may be damage to the hull.

Chocks to support the keel may be solid, swivelling or

roller. The solid ones permit little sliding and one needs plenty of helpers to lift the boat to load her. The swivelling chock under the bow enables the stern of the boat to be swung to, or from, 90 degrees from the trailer centreline. This means that only half the weight is being lifted at a time, the second half, moving the bow to or from the chock being done with the stern of the boat on the ground. This is useful if you can trail close to the water's edge and do not wish to immerse the wheels for a 'trailer launch'.

Roller chocks enable the boat to be rolled on to, or off, the rear of the trailer. This arrangement becomes the more necessary as the size and weight of the boat being carried goes up. This may be coupled with the 'tilt-back' draw-bar, with which the 'roller equipped' end section drops down into the water so that the boat can be drawn on to the trailer by winch. Bilge support chocks generate un-necessary friction to off-and-on loading, and it is useful to drop them down clear of the bilges while it is going on.

Another necessary chock is the bow snubber. This should be mounted on a rigid pillar but is often incor-porated with the mast support. The tendency is for the load to slide forward on a trailer so this item has a job to do, and it should be at the right height to engage the bows just under the gunwale, and be well padded.

Tying down: This is essential and the best investment is a set of Terylene straps, which are quickly attached, kind to the boat and do not shrink when wet. An alterna-tive is the 'strongback'—a piece of timber which straddles the boat and is slightly wider than the beam at the gun-wale. Long hooked rods, each threaded at its top end, engage a welded eye on the trailer and pass through a hole on either side in the strongback, to be tightened by a nut.

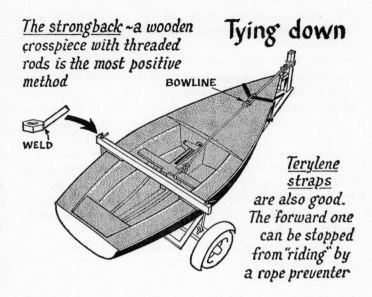

The strongback ~a wooden crosspiece with threaded rods is the most positive method

Tying down

BOWLINE

WELD

Terylene straps are also good. The forward one can be stopped from "riding" by a rope preventer

The Coupling: This is invariably a 50 mm ball, these days, although there is still something to be said for the simplicity and cheapness of the pin type. Maintenance consists of a smear of grease while a plastics cover, when not in use, keeps it (and you) clean. The towing bracket to which it is attached should be custom made for the car, and a number of manufacturers do this.

'Hitch' Height: Since the coupling is not adjustable for height it is useful to have this adjustment at the trailer hitch, which fits the coupling. It will do this if mounted on the pillar by clamps. While level towing is not essential for safety this can be an advantage as it improves the rear view in the interior mirror. There could be an improvement in ground clearance, too.

Towing Characteristics: This must depend as much on

the car as on the equipage and the greater the power/weight over the drag/weight, obviously, the less the tow will be 'felt'.

With non-braked systems the reduction of braking effect must be remembered and considered in one's driving. So too with acceleration, while the extra length when overtaking, and the smaller radius taken by the trailer than the car when turning, are other points to be borne in mind.

Less pleasant characteristics are 'snaking' and 'pitching'. The former consists of the trailer swaying from side to side and causing slight rear end 'drift' to the car. To alleviate this spring adjusting devices are obtainable which fit between coupling and trailer, but a few experiments and checks could be tried out first.

See that the tyre pressures are correct for both the car and the trailer wheels, or experiment with them. It is recommended that, for towing, the car's front wheels are two pounds under normal and the rear ones two pounds over. Another possibility is that the track, or 'toe-in' of the trailer is incorrect, and this could be checked at a garage.

'Pitching' is stimulated by the car's own suspension system. The type of springing, its range and damping all affect the condition. As these are two-wheeled trailers they are bound to rock fore-and-aft and, in bad cases, one must resort to 'spring assisters' for the car. There are various types and it is best to get professional advice as to which to choose for a particular car.

Speed Limits: These are under constant review but, at present, regulations in this country limit the speed of vehicles towing a trailer to 40 mph on unrestricted roads. But if certain conditions are complied with, given in 'The Motor Vehicles (Variation of Speed Limits) Regulations 1973', this figure can be increased to 50 mph.

Among the conditions are that the laden weight of the un-braked trailer must not exceed 60 per cent of that of the towing vehicle and that it carries the figure '50', in white on black, and visible to the rear. The vehicle must have its kerbside weight, including a full tank of petrol but excluding passengers, displayed on its nearside.

Lighting Requirements: A lighting-board, duplicating the registration number and all the lighting requirements of the rear of the towing vehicle, must be carried on the back of the trailer. This does mean 'trailer' officially, but the transom of the boat has gained un-official acceptance.

The number should be similar in letter size, style and colour to that on the car (white on black or black on yellow) and illuminated. There are positional limits for the tail, stop and turn indicator lights so the board needs to be wide enough and mounted at the right height. These, at the time of writing, are between 15 in and 3 ft 6 in from the ground and within 16 in of the extreme widths of the load. Additionally, and within the same limits, two reflective triangles must be carried. These need not be on the board itself, but the sizes are important —equilateral sides measuring between 150 and 200 mm.

To supply the electrics you need a seven-pin plug and socket (although you will not need the auxiliaries pin) and enough multi-core cable to go from the car to the transom. The car-to-socket connections are made in parallel with the car's electrics and this may have the effect of slowing, or quickening, the flashes of the turn indicators beyond the acceptable limits. These are between 60 and 120 flashes a minute. To cure this means the purchase of a 'heavy duty' flasher unit.

Wheels for the Trailer: Wheel failure, whether by

Lighting board

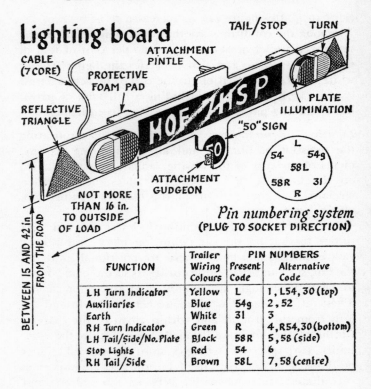

CABLE (7 CORE)
ATTACHMENT PINTLE
PROTECTIVE FOAM PAD
TAIL/STOP
TURN
REFLECTIVE TRIANGLE
PLATE ILLUMINATION
"50" SIGN
ATTACHMENT GUDGEON
NOT MORE THAN 16 in. TO OUTSIDE OF LOAD
BETWEEN 15 AND 42 in. FROM THE ROAD

Pin numbering system
(PLUG TO SOCKET DIRECTION)

FUNCTION	Trailer Wiring Colours	PIN NUMBERS	
		Present Code	Alternative Code
L H Turn Indicator	Yellow	L	1, L54, 30 (top)
Auxiliaries	Blue	54g	2, 52
Earth	White	31	3
R H Turn Indicator	Green	R	4, R54, 30 (bottom)
L H Tail/Side/No. Plate	Black	58R	5, 58 (side)
Stop Lights	Red	54	6
R H Tail/Side	Brown	58L	7, 58 (centre)

puncture or a bearing fault, can be a great nuisance, and a spare wheel is an obvious precaution. To have the trailer wheels the same size as those of the car is one neat way out of this problem. This is sometimes done when the car is a Mini or a small car with similar 520 × 10 tyred wheels—as fitted to the larger dinghy trailer.

But one can reduce the risk of failure by careful attention to tyre wear (which is subject to the same legal minimum tread limits as car tyres are), tyre pressures and lubrication.

The manufacturer's recommendations for tyre pressures are:

2-ply, 25 to 30 psi
4-ply, 30 to 60 psi
6-ply, 45 to 90 psi

Bearings, as the most sensitive parts of the trailer, require the most attention. Whether ball or tapered roller type they have the corrodibility of steel and must be kept well lubricated with a water repellent or a medium viscosity grease.

Most trailer hubs have 'sealed' bearings, which is some protection when the trailer is used as a launching trolley and immersed. But no bearing can be completely sealed and some owners will not use their trailer for this purpose. The cardinal sin is to run the hubs deep into cold water when they are still hot from trailing.

If used for launching, the trailer hubs should be allowed to cool, then given a few pumps of grease before immersion, then given some more after boat recovery. For ordinary use, the hubs should be dismantled, cleaned and packed with grease once a year, then given the grease gun occasionally according to use.

Attention is needed to bearing adjustment, as excess 'play' accelerates wear as much as tightness and binding does.

Sand is the worst thing and care should be taken when the trailer is on the beach, especially to protect the hubs. Salt water is a corrosive and should be washed off with fresh water, not only from the wheels and brake drums, where fitted (they should be flushed out) but also from the framework, even though it be galvanised.

Combination Trailer/Trolley Outfit: If you don't want to use your trailer as a launching trolley, and if you lack the supply of labour to transfer the boat from trailer to a specialised trolley, the 'combination' could be the solution. The boat is mounted on the trolley which, in turn, is mounted on the trailer. Near the water's edge the outfit is unhitched from the car and tilted up at the front. This is enough to allow the trolley wheels to touch the ground, when it is unclipped and separated from the trailer for trolley and boat to be pushed down to and into the water. The reverse procedure is equally simple and allows single-handed management of quite heavy dinghies.

The combined unit is fairly expensive, but not when compared to the price of a standard trailer and trolley purchased separately. In common with the standard 'A' frame trailers, these trailers suffer from not being able to adjust the axle beam for fore-and-aft imbalance. Although minor adjustment may be possible by shifting interior 'ballast' the boat's position may be more difficult to alter. Care should be taken then, to ensure that the outfit chosen is the correct size for your boat.

Launching Trolleys: These are the mobile 'amphibians' which you can push under water with impunity. There are no delicate bearings to worry about; only plain ones which are quite happy to receive the 'second-hand' grease which has already done service in the trailer hubs.

The choice of trolley is decided partly from the size of the boat and partly from the firmness of the approach to the water. Length should be at least half the length of the boat while the width need be no wider than the boat's beam but no less, for the sake of stability than, say, beam minus 12 inches.

The smaller the point of contact each wheel has with the ground the more it is liable to sink in: that is to say, while hard, small diameter wheels are suitable for the

Launching trolleys

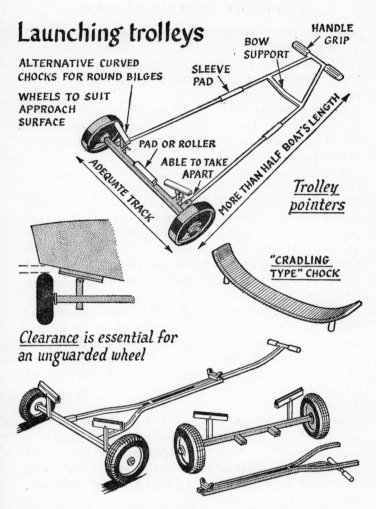

ALTERNATIVE CURVED
CHOCKS FOR ROUND BILGES

WHEELS TO SUIT
APPROACH
SURFACE

BOW
SUPPORT

HANDLE
GRIP

SLEEVE
PAD

ADEQUATE TRACK

PAD OR ROLLER

ABLE TO TAKE
APART

MORE THAN HALF BOAT'S LENGTH

_Trolley
pointers_

"CRADLING
TYPE" CHOCK

Clearance is essential for
an unguarded wheel

A "take apart" trolley stows in the boat or the car for
trailing. This one has a folding handle, too (Beaver Trailers)

hard, metalled, concrete or wooden slipway, for soft ground or sand you must have big diameter 'balloon' tyred wheels. Even so, when the going is really soft no wheel (the type fitted to the moon buggy excepted) will cope.

Other points to watch for in the choice of trolley are the requirement for an adequate clearance between the top of the wheel and the boat's bilge chine, and the type of chocking.

The latter has a bearing on the former, as although one tries to get a low loading height by lowering the adjustable chocks this reduces the clearance. If you can avoid having the short adjustable chocks, do. They have an unpleasant habit of working loose or being knocked askew by the boat and doing damage. The full-width ones hingeing in the centre are better, and they can be adjusted to chine angle, but a solid 'tailored to fit' moulding in resinglass or wood is best of all.

There is advantage in being able to dismantle the trolley for transit, perhaps for stowing in the boat. Some trolley manufacturers have ingenious products which take apart so completely that they may be carried in the boot of the car—assuming you have nothing else in there, of course.

Boat Rollers: When the beach is too soft for the most accommodating wheels another way to get a boat over is by boat rollers. These long, inflatable 'sausages' are equally able to carry a boat over unyielding stones and rocks, over which wheels are also 'at a loss', although there may be more risk to wear and tear.

Progress is slow but fairly effortless and not less than two are needed—although the more, the better. With more rollers the boat can be kept level, whereas with two she may have to be lifted as each roller is taken from

The beach

*The softer
the approach;
the wider
and larger
the wheel
preferred*

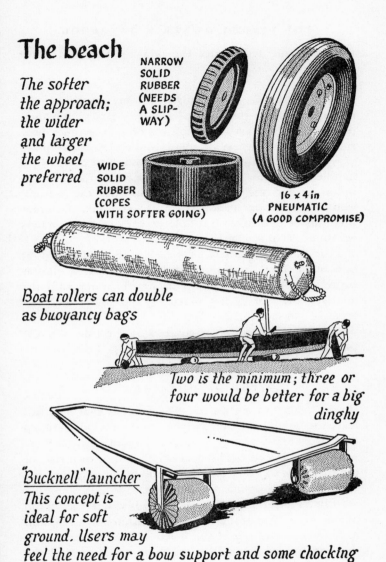

NARROW
SOLID
RUBBER
(NEEDS
A SLIP-
WAY)

WIDE
SOLID
RUBBER
(COPES
WITH SOFTER GOING)

16 x 4 in
PNEUMATIC
(A GOOD COMPROMISE)

<u>Boat rollers</u> can double
as buoyancy bags

*Two is the minimum; three or
four would be better for a big
dinghy*

<u>"Bucknell" launcher</u>
*This concept is
ideal for soft
ground. Users may
feel the need for a bow support and some chocking*

the back and 'fed' under the bow for the next move forward.

'Hybrid' Launcher: I apologise for my epithet to the manufacturers of this type of launcher (Barry Bucknell designed one) which seems to be a cross between a boat roller and a trolley. Wide, plastics drums take the place of wheels which are mounted on an axle and pulled by a handle. The idea seems good, especially for soft approaches, and the possible criticism—that the 'wheels' would float on immersion is countered by the fact that they can be drilled with a hole or two (to admit water) without loss of strength.

This buoyancy objection has also been levelled at pneumatic tyres for launching trolleys with a similar suggested solution—to pump in a little water with the air. Actually this is not much of a problem and can even be an advantage when the trolley is accidently dropped in deep water or covered by the tide, when it partially floats.

THE DINGHY PARK

If you belong to a club the likelihood is that you will leave the boat in the dinghy park for the periods between outings and have a more or less permanent position there —no doubt at a fee! The boat is left at your own risk, or the insurer's, so every effort must be made to ensure her security and minimum deterioration.

Insurers tell us that when a dinghy is so left, she should not be on a road trailer, as this facilitates stealing, and to remove the wheels if she is left. This last measure is so extreme that I would consider it easier to trail the boat home each time—but, there it is.

If left in a high 'theft risk' area, certainly nothing

portable of any value should be left in the boat. Sail, rudder, anchor, tiller, oars and rowlocks come into this category.

Then the boat should be chocked and tied down, not to prevent her being stolen, this won't stop the determined thief, but to prevent her blowing over in high winds and damaging herself and others.

The equipment for doing this I call, for want of a better name, 'dinghy park furniture'. These consist of single or double chocks (to support the bilges and prevent 'rocking'), a prop (for the bow, to facilitate drainage from the transom ports) and concrete 'anchorage weights', stakes or 'ground corkscrews' (to give the secure anchorage points to which to tie the boat down). Smaller chocks, or convenient sized stones, are also required for chocking the wheels.

Further description is unnecessary, as a visit to any dinghy park will produce enough ideas, and examples to set anyone to work knocking-up items with wood, padding, rod and concrete.

THE BOAT COVER

This item may perhaps be omitted from your 'extras' list if you have a resinglass, polyethylene, EPS, or aluminium boat, or even a wooden one, if she is kept well painted and varnished. Yet the extremes of weather in this country, from ice to blazing sunshine, are such that an insulating cover does give extra life to *all* boat material constructions. But there is another reason for having one—cleanliness.

After the boat has been brought ashore she needs a wash down, especially if the sailing water is salt. If a hose is available this enables the centreboard case to get a wash too, otherwise it means a sluice from the bucket with the dinghy's transom ports open and while inclined on the slipway.

A leather-down and she is clean and ready for next time—but only if a cover is used. Without it the boat will be bestrewn with leaves, sand or dirt when you come back next week.

Choice of design for a cover is between the 'overboom' and the 'flat'. The former is a 'shedding' cover in that the rain stands more chance of running off, and not collecting in puddles to form 'birdbaths', as is the case with the flat cover—unless the boat is well tilted.

Material choice rests between the traditional canvas, PVC coated nylon, reinforced polythene and a flexible fibreglass. Canvas, which should be at least 12 oz, and preferably 15 oz should be good for four or five seasons if it is proofed annually. Its virtue is that it 'breathes' and allows little condensation to build up. There is a shrinkage problem but manufacturers know enough about their material to make covers adequately oversize to allow for this.

The 'man-made' materials, being impervious, have condensation problems to a greater or lesser degree. PVC-nylon, with its absorbent lining, is the best of them.

From a price point of view there is little object in saying anything. Canvas was always the most expensive and suffered a doubling of its price during the year before writing. Yet man-made materials, especially oil-derivatives, are jumping in price too. The current market prices are what matter to you, and me!

But man-made ropes are the best for under-boat ties, at whatever price. They combine strength with non-shrinking.

LAYING-UP

As mentioned elsewhere, the dinghy sailing season is

never really finished, and for those who can sail inland and who have the right clothing, good blood circulation and enough enthusiasm, the winter is as good as, if not better than, the summer.

For the rest of us November or early December sees us taking down the burgee and packing the boat away for her winter rest. Wooden boats especially will last longer, or can be maintained easier, if protected by interior storage from the ravages of ice and frost.

The first job is to give the boat a really good wash down, as even under cover, dirt, or salt will attract damp from the air. The centreboard can be removed and the case flushed out. As many items as can easily be removed and stored at home, should be taken off. This includes buoyancy bags (which can be partially deflated and stored unfolded) and fittings, which would have to be removed before varnishing next year anyway. These can be cleaned and greased as a winter task. Covers, or bungs in buoyancy tanks also need to be removed as part of the grand strategy of allowing as much air to circulate as possible for drying out and the discouragement of rot.

Undercover storage for the hull, especially if she is wooden, should now be sought. Lucky are those who have spare garage room, either at ground level, or at height for the boat to be suspended above the car. Some clubs rent boat-shed space to their members while some boatyards may have storage capacity.

If the boat really must winter outside try to arrange a sort of tent cover over her, instead of the normal boat cover, which does not allow sufficient ventilation.

The mast may have to be stored outside and, in its horizontal position it needs to be supported at the minimum of three points, all exactly in line, to avoid distortion. Forestay and shrouds can be removed and stored and if the latter are labelled 'P' and 'S', this will help fitting out. The same applies to pairs of fittings from the hull (eg. fairleads with their 'own' screws). Halyards too, can be taken and stored away from the winter's grime, but if these are internal it may save a fiddly job next year if a length of nylon line, or similar, is drawn through as they are pulled out, to 'pilot' them through again.

RESINGLASS MAINTENANCE

Despite what the advertisers say, resinglass is maintenance-free only if you are prepared to accept a hull with digs and scratches, 'crazing' of the gel coat and a lacklustre appearance. The gel coat is the waterproof skin and if this is penetrated there will be some water absorption, varying with the density of the lay-up. If this is followed by freezing, there will be expansion, cracking, and further deterioration.

The areas to be painted must be thoroughly degreased from the wax 'release agent' used in manufacture. This is done by scrubbing with detergent and/or a proprietary oil-removing fluid. Do not use liquids such as carbon-tetrachloride, acetone or trichlor-ethylene which would damage the surface. When the wax is cleared—and the test is if water wets the surface evenly rather than going in puddles and rivulets—the area can be rubbed down with 240 grade wet-and-dry paper, used wet. A good 'key' must be obtained for successful paint adhesion.

Deep indentations can be made good at this stage, using an epoxy stopper or filler paste from a repair kit. This is a two-part material which is mixed just before use and 'dries' chemically. These repairs must also be rubbed down before proceeding. Take care not to re-grease the hull with finger-marks.

A coat of resinglass primer, (which is a polyurethane two-pack paint that also has to be mixed before application) goes on next, followed by one, two or more coats of the chosen coloured two-pack polyurethane, until the desired finish is obtained. Thinned epoxy filler or trowel cement may be used between paint coats for 'facing up' and care should be taken that the maximum recoating period between coats (to ensure bonding) is not exceeded. This is 36 hours for polyurethane. Failure to do this means rubbing down again to obtain a mechanical key.

If a non-slip surface is required (for the floor or part of the side-decking) silver sand should be sprinkled on the penultimate coat, when wet (using an old stocking as a sprinkler), and the final coat applied when dry.

WOODEN BOAT MAINTENANCE

Wooden boats can have a long life if they are never allowed to deteriorate. A season or two's neglect can make the repainting job a hard one, and store up trouble for years to come. Regular maintenance makes the job easy. Particular care should be taken to keep the end grain of plywood, if it is exposed, well sealed with paint or varnish, for this is the starting point of delamination. Scrapes and scars acquired during the season should also be touched up as they occur, a small tin of primer and varnish being carried for the purpose.

There is a choice to be made between two paint systems —'high performance' and 'conventional'. Polyurethane (already mentioned under resinglass) and epoxy paint make up the former. They are entirely synthetic-resin based and dry, or cure, chemically. They are completely impervious to water and very hard, but depend for success on a good initial bond to the substrate (the bare hull), on the continuance through its working life of that bond (it can be undermined by damp passing through a porous substrate, eg. wood) and on the maintenance of an unbroken surface (another entry for damp). Failure on any of these points leads to flaking off.

Conventional marine paint is, with research and development, a far cry from the traditional oil-bound paint it suggests. Acrylic and 'one-pot' polyurethane come into this category and they all contain natural oils as well as synthetic resin and dry by contact with the air. They are not entirely impervious to water and are more tolerant to a substrate with moisture content. A well painted boat can actually 'dry out' during a spell of very dry weather, and this slight porosity spells a possible trouble source if a substrate has a conventional system on one side and high performance on the other. But despite its greater tolerance, conventional paint still requires care in preparation, application and overcoating, as will be described.

Choice between these systems is a personal one and may even be settled by cost (high performance is dearer) but it is easier to change from high performance to conventional, if dissatisfied, than vice versa. In fact, polyurethane *must* not overcoat conventional or the base paint will wrinkle. To change, all the old paint would need to be stripped off and the bare surface thoroughly sanded.

Non-compatability also applies if the systems are merely abutted; edge wrinkling will occur if the high performance is put on after the conventional. There are two systems in varnishes as well as in paint and the same stipulations apply.

Preparation of Surface: If this is new wood a good sanding is required before painting, particularly if it is plywood, as the surfaces of the veneers are hard and almost 'glazed' from the presses. Rub down with 200–280 glasspaper, rubbing across the grain if this is to be painted and along it if it is to be varnished (the scratches would show through) then finish with garnet paper. Dust can then be removed first by vacuum cleaner, then with a rag soaked in thinners and the surface allowed to dry.

If this is old paint or varnish work to be given a complete repaint, all the old paint or varnish must be removed and the surface treated as for new wood.

Paint is removed by burning and scraping, the use of a power sander, or disc, or the use of chemical stripper. Varnish work debars the first two methods, as burns and scratches cannot be hidden, and we are left with the chemical stripper and scrapers.

Old paint or varnish which is in sound condition needs only a good rub down to provide a key for the new applications. Use 240 grade wet-and-dry used wet with plenty of water and a little soap to aid lubrication. Wash well, and dry.

Stoppers and Fillers: These marine-type putties are expected to perform miracles of restoration—and they do, if used correctly.

Epoxy stopper or polyester paste may be used under paintwork; it is gap-filling and suffers no shrinkage. Under varnishwork a 'plastic wood' (made up from a gap-filling resin glue, like Cascamite, and into which a quantity of sawdust of a matching colour is mixed) can achieve a similar effect.

The trowel fillers and quick-drying stoppers are air dried and must be used only between paint films, as bare wood will absorb their solvents and leave them weak and powdery. There is some shrinkage, and deep holes should be filled in stages. They should then be faced off and sealed with undercoat or primer before proceeding with the next coat.

Thinners: Use only the thinners recommended and sold

by the paint manufacturers for thinning paint. There are types of thinners for epoxy, polyurethane and conventional paints and these are not interchangeable.

High-Performance Paint Systems for Wood: If this is to be the complete system prepare the bare wood as detailed above then give one coat of primer. This will be either an epoxy primer or a penetrating wood primer. The former cures hard but the latter does not completely dry until the first coat of the chosen polyurethane enamel is applied. Filling and facing up can now be done, using a stripping knife to feed the trowel cement into the depressions and, preferably, leaving a day to harden before rubbing smooth.

Further coats of polyurethane can now be applied, bearing in mind the recoating period during which bonding to the previous coat is assured, with little or no rubbing down, until the required finish is gained. Four coats would be a reasonable number.

For normal repainting rub the surface down with grade 240 wet-and-dry used wet and go straight on with the stopping and facing, followed by two coats of polyurethane.

Conventional Paint Systems for Wood: Prepare the bare wood, as usual and prime with a conventional primer or an epoxy primer. When dry, make up the surface and follow with another coat of primer. Rub this one down lightly with wet-and-dry and make up the surface again, if necessary. Then apply two coats of conventional undercoat, of the appropriate colour, sanding lightly after each and then the final top-coat enamel.

For a normal repaint, just rub down with grade 240 wet-and-dry, wash, dry and do the facing up, then apply one coat of undercoat and one of enamel.

Varnish: Unlike paint, varnish can tell no 'lies' for the bonding of the underlying coats and the condition of the wood itself are all revealed.

Bare wood should be rubbed down as for paint, the strokes going with the grain instead of across. The wood must be clean and dry; use the vacuum cleaner, brush and wipe with a rag soaked in thinners, each in turn to remove dust.

If conventional varnish is being used, dilute the first coat with 20 per cent thinners and brush well in—or rub on with a fluffless rag. Apply the second and third coats at 24-hour intervals (two coats a day can be put on if using one-pot polyurethane or moisture-dried varnish, which is a quick dryer) without rubbing down. Then rub down lightly with 320 grade wet-and-dry used wet and follow with two or three more coats, lightly rubbing down with 320 between coats, this time used dry. This is just 'de-nibbing' the surface.

For a really first-class job, leave this four weeks for the varnish to cure. By this time the surface will appear corrugated, as the varnish will have shrunk and taken up the undulations of the grain. So rub it down with grade 240 used wet. This flattens it but leaves it glazed, so dry the surface and rub it down again with grade 320 wet-and-dry, used dry. Then follow with two more coats with de-nibbing between. This final treatment is suitable for revarnishing work.

For polyurethane varnish, the bare wood should first be sealed with a polyurethane varnish primer, a wood primer or the varnish thinned with 10 per cent polyurethane thinners. The build up of coats can then proceed as for conventional varnish, care being taken to re-coat within the period for doing so, (between 6 and about 36 hours) otherwise there will be more rubbing down to do for a mechanical key.

Burnishing: This is suitable for polyurethane varnish but not for conventional. The varnish should be left to cure, which takes a week to ten days at normal temperatures, then rubbed with fine grade wet-and-dry, used wet, to remove blemishes. It is then polished with an abrasive car polish, and maintained with applications of the same during the season.

Sealing End-Grain and Weathered Surfaces: Particular care must be taken to seal the end grains of plywood with thinned varnish, as this is where delamination sets in. This also applies to solid wood, such as the ends of gunwales, thwarts, and the ends of the planking on clinker boats.

Varnish can be only as good as the wood underneath it. Weathered surfaces, with cracked and de-natured grain, will repeat their imperfections in the varnish film. The choice then is whether to paint it, starting the bare wood off with a metallic wood primer, or to give it the linseed oil treatment. The latter is suitable for subsequent varnishing or painting, but does take a lot of time.

You must buy boiled linseed oil. It is brushed or rubbed on to the bare wood and left to soak in and swell the grain. It may take two weeks or more before this is absorbed, after which more can be applied, and it will take a similar period. Further 'coats' may be put on for the wood to be completely sealed, then the surplus is wiped off with white spirit and left to dry. After that the varnish, which must be conventional, can be applied in the normal way.

APPLICATION OF PAINT AND VARNISH

Speed and evenness are the things to aim for and the large expanses of hull and decks are the most difficult.

Progress by starting at one end and doing a section, say from keel to gunwale if this is the hull, spreading the paint, brushing it out vertically and 'laying-off' horizontally. Then apply the next section alongside while the former is still wet enough to 'blend' the one into the other when laying-off. As paint or varnish soon loses its 'brushability', this ability to blend decides the width of each section.

Matters bearing on this are: the consistency of the paint (it can be thinned with the appropriate thinners in cold weather, or it can be warmed); the ambient atmosphere (a temperature of 65 degrees is ideal, and neither too damp nor too dry); wind (which may surface-dry the paint very quickly, so there is an additional advantage in working inside, if possible); the size of brush (which should be as large as the intricacies of the work permit); and the skill of the painter.

Paint should not be applied too thick, or it will sag in 'curtains', but varnish can be thicker. Brush angle should be steeper for spreading than for laying-off when only the tip of the brush is used. Paint must be well stirred but varnish is left unstirred (at least for a period before use), to eliminate air bubbles.

Brushes: A range of sizes of $\frac{1}{2}$ inch, for patching and cutting waterlines, $1\frac{1}{2}$ or 2 inches, for interior work and 3 or 4 inches, for decking and topsides, is required. Varnish and conventional paint brushes need more 'body' to them than polyurethane brushes do.

They must be kept clean and I buy gallon cans of cheap white spirit for the purpose. Half a dozen rinses in a tin lid, then a rub with detergent and a rinse in water, cleans a brush if it is done immediately after use. Varnish brushes can be kept hanging in a jar of white spirit from day to day if kept in use, for up to a week. Polyurethane

thinners must be used to clean brushes if that paint has
been used.

Appendix:

Useful References

British Standards Institution, 2 Park Street, London W.1
Dinghy Cruising Association, 33 Blythe Hill Lane,
London SE6 4UP
National Sailing Centre, Arctic Road, West Cowes, Isle
of Wight
National Schools Sailing Association, Education Office,
County Hall, Chichester, Sussex (see page 20)
Outward Bound Trust, 34 Broadway, London S.W.1
Royal Society for the Prevention of Accidents, Royal Oak
Centre, Brighton Road, Purley, Surrey CR2 2UR
Royal Yachting Association, Victoria Way, Woking,
Surrey (see page 20)
Sail Training Association, No. 3 Glencoe, Bosham Lane,
Old Bosham, Chichester, Sussex
Ship and Boat Builders National Federation, 31 Great
Queen Street, London W.C.2 (see pages 160, 164)

For up-to-date information on all the material needs,
services, etc. which the dinghy-owner requires, a
thoroughly useful reference book is *Boat World*, published annually by Sell's Publications Ltd.

Diagrams for Easy Reference

Anchors, 210

Blocks, Sheet, 114

Courses, The, 204

Depression, Chart of a typical, 188

Hull sections, 39

Knots, bends and hitches, 98

Launching trolleys, 257

Launching rollers, 259

Mast sections, 104

Racing: Rights of way, 220

Racing: Keeping clear, 223

Racing: Techniques, 233

Reefing: Mainsail, 146

Reefing: Gunter mainsail, 148

Rigs, 45

Righting a capsize, 171

Rudder and tiller, 137

Sail parts and definitions, 72

Sail folding, 80

Sail patching, 82

Sail controls, 86

Sail balance, 132

Sculling, 181

Shackles, 112

Spinnaker gear, 237

Splicing and whipping, 94

Steering by rudder and by sails, 131

Tackle applied to mainsheet systems, 119

Trailers, 248

Trailer lighting board, 254

Vector diagram: apparent wind, 194

Vector diagram: surface wind relative to current, 196

Tables for Easy Reference

Accessories for dinghy, 64

Anchor and cable dimensions, 211

Beaufort wind scale, 186

Buoyancy, Personal: Minimum requirements, 160

Buoyancy, Unit: Supporting ability, 166

Cost: Dinghy-owner's annual outlay, 26

Rod wire breaking stresses, 95

Ropes, Natural and synthetic, 95

Ropes, Choice and size, 99

Sailcloth Weight Conversion, 100

Shackle, 'D': Breaking loads, 111

Shackle, Snap: Breaking loads, 112

Signals, Flag (Racing), 242

Wire rope breaking stresses: standing rigging, 93

Wire rope breaking stresses: running rigging, 95

Index

(Diagrams and Tables for easy reference
listed on pages 275-6.)
Entries in SMALL CAPS denote Classes of dinghy. Page
references in *italics* denote illustrated matter.

ABS hulls, 34
Aluminium hulls, 35, 164
Anchoring, 209ff, *210, 211*
Anchor weight, 211
Aspect Ratio (A/R) of
    sail, *74, 75,* 144
    of rudder, 136, *137*

Backing wind, 187
Bailers, Self-, *176,* 177
Bailing, 174ff, *178*
Barber hauler, *86,* 90
Battens, *45,* 49, 83
BEACHCOMBER, 53
Beam factor of hull, 37
Beaufort wind scale, 186
Bermudian mainsail, 83
    Setting, 87
    rig, 44, *45,* 47
Bilge pump, 177, *178*
    runners, 40
Blocks, 113ff, *114*

BOBBIN, 58
Boom, 101, 145
Bowsprit, 101, 108
Breeze, Sea, 189ff, *192*
Bumkin, 46, 108
Buoyancy, Hull, 33, 35, 36,
    41, 42, 163ff, *165*
    Mast, 102
    Personal, 156ff, *158*
    aid, 158ff

CADET, INTER-
    NATIONAL, 21, 22
Capsize, 149ff, 168ff
Cat rig, *45,* 47
Catamarans, 49, 53, *54*
Centre of Buoyancy (CB),
    138
    of Effort (CE), 131ff,
    143ff, *144*
    of Lateral Resistance
    (CLR), 131ff

Centreboard, 43, 69, 123ff, *126*, 134, 174, 179, 206, 209

Certificate, Class Measurement, 69, 215

Children and sailing, 20ff, *59*, 156ff

Chine construction, 30ff, *31*

Class racing, 215

Clinker-built hulls, Wooden, 27ff, *28*, 56

Clinker, Simulated, 56, 58

Clothing for sailing, 149ff, 162ff

Clouds and wind, *188*, 189, *192*, 193

Clubs, Sailing, 16ff, 182

Cold, Dangers of, 161ff

Colour of ropes, 96
of clothing, 155

CONTENDER, INTER-NATIONAL, 53

Cost of sailing, 22ff, 64ff

Courses, points of sailing, 203ff, *204*

Crew number, 52
weight, 140ff, *141*, 205

Cruising dinghies, 55ff

Cunningham hole gear, 83, 84, 88

Current, 195ff, *196*, *199*, 234

Cutter rig, 44

DABBER, 56, 57, 149

Daggerboard, 43, 61, 125ff, 134

Decking, 40, *41*, 42ff

DEVON YAWL, 58

Dinghy park, 260

Do-it-yourself boatbuilding, 66ff

Drag, 38, 125, 129, 136

Drainage, Hull, 174ff, *175*

Draining hull, Self-, 33, 61, 164ff, *165*, *175*

Drascombe-built dinghies, 56, *57*, 60

Drop-plates, 125ff, *126*

ENTERPRISE, 11, *12*, 52, 245

EPS (expanded polystyrene) hulls, 34ff, 68, 164

Exposure, *see* Cold

Fairleads, 76

FELIX, 53

Fenders, 213

FINN, INTER-NATIONAL, 53, 106

FIREBALL, 40, 245

Fishing, 57, 61

5-0-5, 52

Flags, Code, 216, 217, 226, 242ff

Flares, 183

Flow control, Sail, *84*

Footwear, 151ff, *152*, 163
Foredeck, 42
FORELAND, 61
Foresails, 78, 81, *84*
   Setting, 85
   Reefing, 147
FOURTEEN, INTER-
   NATIONAL, 40, 62
Freeboard, 38
Friction layer, Atmos-
   pheric, 189
Furling luff spar, 147

Gaffer rig, *45*
Gate start, 216ff, *218*
Gooseneck downhaul, 83,
   89, *122*, 145
'Goosewinged' sails, 47,
   205
GP 14, *14*, 52, 58, 64
GRP (glass-reinforced
   plastic) hulls, 32ff,
   56, 66, 69, 163ff
   Double-skinned, 33ff
   maintenance, 264ff
GULL, *59*, 60
Gunter mainsail, 84
   Setting, 90
   rig, 43ff, *45*, 103
   Reefing, *148*
Gusts, *190*, 191

Halyards, 93, 99, 236, 264
Handicap racing, 215
Headgear, *152*, 153, 163

Heaving-to, 209
Heel, Angle of, 38, 128, **203**
Heeling lever, 138ff, 143
   moment, 138ff, *139*, 143ff
HERON, 60
Highfield lever, 85, *86*, 90
HORNET, 52, 142
Hull construction and
   materials, 27ff
   Displacement, 36ff, *37*
   Planing, 36ff, *37*
   sections, *39*
   shape, 36ff, *139*

Jack Holt flow control, **83**
Jib furling gear, 148
   stick, 110
JOLLYBOAT, 52
Jury rig, 179

Keel, 40
Ketch rig, 47
Kicking strap, 87ff, 100,
   117, 121, *122*
Kits, Building from, 67
Knots, bends, hitches, 97ff,
   98

Lapstrake hulls, 56
Launching and landing in
   current, 198
   in waves, **202**
   rollers, 258ff, *259*
   trolley, *246*, *248*, 255ff,
   257

Laying-up, 261ff
Lee-bowing, 198ff, *199*
Lee helm, 130ff, 144
Leeway, 128ff, *129*
Lifejacket, 157ff
Light-weather sailing, 208, 232
Line start, 216, *218*, 228ff
LONGBOAT, 56
LUGGER, 56
Lugsail rig, *45*, 47ff, 103, 108
LYMINGTON SCOW, 11

Mainsheet systems, 118ff, *119*
Marks, Rounding and giving room at, 222ff
Mast, 49ff, 101ff, 134, 263
    Aluminium alloy, 101
    Bamboo, 103
    bend, 88, 101ff
    Carbon fibre, 103
    rake, 76, 91
    Rotating, 105ff, *107*
    sections, 103ff, *104*
    Stainless steel, 102
    stepping, 106ff, *109*
    Two-part, 106
    Unstayed, 49, 106
    Wooden, 101
Materials for dinghy construction, 27ff
MERLIN ROCKET, *15*, 62, 75

Methods of dinghy construction, 27ff
MINISAIL, 53
MIRROR 10 and 16, 40
    58, 111, *119*, 121, 245
Mizzen mast, 46ff
    sail, 46ff
Mooring, 211ff, *212*
MOTH, INTER-NATIONAL, 53, 62, *63*, 142, 149
Multihulls, 53

NAIAD, ESB, 61
National Schools Sailing Association (NSSA), 20
NATIONAL 12, 62, *63*
NORFOLK, *54*, 62

Oars, 60, 179ff, *181*
OK, 106
One-design class, *54*, 57, 61ff
OPTIMIST, 49
Outboards, 60

Paddles, 60, 179ff, *181*, 241
Painters, 100
Paints, Marine, 265ff
Parallelogram of forces, *124*
Pathfinder dinghy, 217ff, *218*
Penalties for infringements, 224ff, *225*

Planing, *199*, 206ff
Plywood hulls, Moulded, 28ff, *29*, 67
  Marine, Chine construction, 30ff
Polyethylene hulls, 34
Portsmouth Numbering System, 215

Racing dinghies, Choice of, 52ff
  finish, 242
  rules and techniques, 214ff, *233*
  starts, 216ff, *218*, 228ff, *230*
Rake of mast, 76
Reefing, 144ff, *146*, 187
  claw, 145
Resinglass, *see* GRP
Restricted class, 61ff, *63*
Rig, 43ff, *45*
  adjustments and tuning, 91ff, *91*
  Hard and soft, 50
  loadings, 113
Right-of-way when racing, 219ff, *220*, *223*
Righting capsized dinghy, 170ff, *171*
Righting lever, 139ff
  moment, *139*, 140
Rod wire, 95
Roll tack, 208, 241
Roller reefing gear, 147

Roof-rack as dinghy transporter, 244ff, *246*
Rope colours, 96
  Natural and synthetic, 95ff
  sizes, 99ff
  and see wire rope
Rot, Checking for, 69
Rowing, 60
Royal Yachting Association (RYA), 20
Rubber craft, Inflatable, 36
Rudder, 130ff, 135ff, *137*, 179, *181*, 209

Safety afloat, 149ff
Sailing areas, Choice of, 12ff
Sail, 71ff, *72*
  balance, 130, *132*
  care, 80ff, *82*
  cloth, 78, 100, 235
  controls, *86*, 88
  cuts, *79*
  folding, *80*, 81
  Fullness and flatness of, 73
  shape, 73ff, *74*
  and see Foresail, Mizzen sail, Spinnaker
Sculling, 180ff, *181*
SEAFARER, 58
Seating, 42
Second-hand, Buying, 68

Self-rescue, 149ff, 168ff
Shackles, 111ff, *112*
SHARK, ANDERSON, 55
SHEARWATER, PROUT, 53
Sheets, 99
   and see Mainsheet systems
Shrouds, 77ff, 108
Signals, Emergency, 182
   Racing, 216ff, 242ff
   and see Flags, Code
SKIPPER, 60
Sloop rig, 44, *45*, 76, 77
Slot effect in sails, 76ff, 77
SMALL CRAFT
   SHETLANDER, 60
SOLO, 53
Spars, 101ff
   Aluminium alloy, 102
   Wooden, 101
Spinnaker, 235ff, *237*, *239*
   boom, 109, *110*, 236ff, *237*
   chute, 236ff, *237*
   guys, 99, 236, *237*
Splicing, *94*, 97
Spreaders, 50, 88, *89*
Springs, Mooring, 213
Spritsail rig, *45*, 49
Steering, 130ff, *131*
Stem profile, 39
Stern profile, 39
SUNFISH, 60
SUSSEX COB, 60

SWIFT, 53, *54*

Tackle, Block and, 117ff, *119*
Tidal flow, 195ff, 234
Tiller, 135ff, *137*
Toestraps, 140ff, *141*
TOPPER, *59*, 60
Tows, Accepting, 183
Trailers, 245ff, *248*, *251*, *254*
Trampolene, 142
Trapeze, *141*, 142ff
Trim, Hull, 135
Trimaran, 55

Underwater hull sections, 38ff
   profile, 39
UNICORN, 55

Varnish, 267, 269ff
Veering wind, 187, *190*

WATERCAT, 53
Waterproof clothing, 153ff, *155*
Waves, *199*, 200ff
WAYFARER, 52, 55, 57, 149
Weather forecasts, 186
   helm, 130ff, 136
Wet suits, *150*, 151, 162
Wetted area, 125

Whipping, *94*, 97

Wind, Apparent, 193, *194*
Clear (for racing), 229ff
Convected, *192*
Gradient, 187
and sailing, 185ff
Sailing techniques in strong, 203ff
Surface, 194ff, *196*
Thermal, 189, 191
True, 71, 194
and see Light-weather sailing

Windage, Hull, 164

Wire rope, 92ff, *94*

Wooden boat maintenance, 265ff

YACHTING WORLD DAYBOAT, 58

Yards, 101

Yawl, *45*, 46, 108